Biofuel Technologies for a Sustainable Future: India and Beyond

RIVER PUBLISHERS SERIES IN ENERGY SUSTAINABILITY AND EFFICIENCY

Series Editor:

PEDRAM ASEF
Lecturer (Asst. Prof.) in Automotive Engineering,
University of Hertfordshire, UK

The "River Publishers Series in Sustainability and Efficiency" is a series of comprehensive academic and professional books which focus on theory and applications in sustainable and efficient energy solutions. The books serve as a multi-disciplinary resource linking sustainable energy and society, fulfilling the rapidly growing worldwide interest in energy solutions. All fields of possible sustainable energy solutions and applications are addressed, not only from a technical point of view, but also from economic, social, political, and financial aspects. Books published in the series include research monographs, edited volumes, handbooks and textbooks. They provide professionals, researchers, educators, and advanced students in the field with an invaluable insight into the latest research and developments.

Topics covered in the series include, but are not limited to:

- Sustainable energy development and management
- Alternate and renewable energies
- Energy conservation
- Energy efficiency
- Carbon reduction
- Environment

For a list of other books in this series, visit www.riverpublishers.com

Biofuel Technologies for a Sustainable Future: India and Beyond

Editors

Yashvir Singh

Graphic Era Deemed to be university, Dehrdaun, India

Universiti Tun Hussein Onn Malaysia, Parit Raja, Batu Pahat Malaysia

Prateek Negi

Graphic Era Deemed to be University, Dehradun, India

Wei Hsin Chen

National Cheng Kung University, Tainan, Taiwan

River Publishers

Routledge
Taylor & Francis Group
LONDON AND NEW YORK

Published 2023 by River Publishers
River Publishers
Alsbjergvej 10, 9260 Gistrup, Denmark
www.riverpublishers.com

Distributed exclusively by Routledge
4 Park Square, Milton Park, Abingdon, Oxon OX14 4RN
605 Third Avenue, New York, NY 10017, USA

Biofuel Technologies for a Sustainable Future: India and Beyond / by Yashvir Singh, Prateek Negi, Wei Hsin Chen.

Routledge is an imprint of the Taylor & Francis Group, an informa business

ISBN 978-87-7022-634-9 (print)
ISBN 978-10-0079-520-2 (online)
ISBN 978-1-003-33832-1 (ebook master)

Contents

Preface

Biofuel is becoming the focus of public and scientific debate due to the severe environmental damage caused by fossil fuels. Biofuel is a resource that is considered a renewable source having a major contribution in various areas and may have several applications in different sectors. The usage of biofuels has grown significantly, with renewables now accounting for a significant contribution to global generation. However, eliminating all fossil fuel energy sources is becoming increasingly vital as otherwise there will be severe environmental and public health consequences. Our book 'Biofuel Technologies for a Sustainable Future in India and Beyond' investigates energy production, elective and inexhaustible fuels, and approaches to assemble a sustainable energy future. This book addresses the need for biofuels, especially in India. This book will provide knowledge on biofuel applications, feedstocks available for biofuel production, and challenges related to their generation processes. Also, the role of nanomaterials as an additive to biofuel is presented in this book. The book stresses the significance of various engineering disciplines and the complete shifts from conventional to non-conventional resources while considering the environmental challenges. The chapters cover different points of view concerning biofuels advances that are seldom discussed in major conferences and journals, which are related to design techniques, innovation, and technological advancement. This book blends innovative, unpublished, and very elegantly compiled findings and progress in the field of "Sustainable Future" and has a strong significance for educational foundations, professionals, analysts, and industry players.

Chapter 1 examines India's renewable energy resources and makes new efforts to describe their availability, current state, significant achievements, and future potential. Additionally, this study examines possible legislative solutions for eliminating barriers and expediting future renewable energy adoption.

Chapter 2 provides a solution to the problem associated with industries that discharge water into river bodies, increasing water pollution and causing severe damage to humanity. The issue can be resolved by utilizing waste water to grow plants. In this context, an experimental study has been carried out to see the effect of industrial waste water on the growth of seasonal plants readily available in the market.

Chapter 3 outlines the increased demand for nanotechnology vis-à-vis the energy domain and focuses on the different areas of energy in which nanotechnology has made a significant impact. Further, it is followed by a discussion on the toxicity effects of the conventional synthesis methods employed. Also, the fabrication of nanomaterials onto different energy devices, and lastly, the fate of such nanomaterials during the consumerism of such devices.

Chapter 4 describes the investigation of the catalytic cracking of waste cooking oil using different activated carbon-supported transition metal oxide as a catalyst to understand better the relationship between catalyst properties and product yields and selectivity.

Chapter 5 reveals the formation of biomass with the help of land, water, solar energy, and associated ways. Edible resources are not yet a viable option in getting biofuels. In India, especially Jatropha, Polanga, and Karanja, such biofuels are preferred under the "National policy on Biofuels" to avoid the food crunch by using edible feedstock.

Chapter 6 presents recent research and development on the applications of mono and hybrid nanofluids for different solar energy technologies and challenges with nanofluids on utilizing in solar energy systems. Finally, based on the literature survey, some suggestions for future work are discussed.

Chapter 7 comprises an overview of vegetable oil feedstock, the effect of direct use of vegetable oil in diesel engines, biodiesel derived from vegetable oil, and the current status of biodiesel in India.

Chapter 8 includes possible enhancements to conventional Internal Combustion (IC) Engines and futuristic technologies driven by sustainable energy sources like The Green Engine, Hybrid Electric vehicles (HEVs), and Hydrogen as a potential fuel for combustion engines (H_2-ICEs).

Chapter 9 provides an overview and references for technical parameters and characteristics of various disposal and utilization methods via thermochemical conversions (combustion, gasification, and pyrolysis) for phytoremediation plants contaminated with HMs. The reduction rate of plant biomass with HMs is high using combustion, gasification, or pyrolysis. Solid, liquid, and gas products produced in the treatment process can be used as resources.

List of Figures

List of Tables

List of Contributors

Acharya, Ankit, *Undergraduate, School of Mechanical Engineering, KIIT (DU), Bhubaneswar, Orrisa, India; E-mail: acharya2000ankit@gmail.com*

Agrawal, Madhu, *Department of Chemical Engineering, Malaviya National Institute of Technology Jaipur 302017, India; E-mail: madhunaresh@gmail.com*

Bhushan, Brij, *Department of Chemistry, Graphic Era University, Dehradun, Uttarakhand, India; E-mail: brij.chem@geu.ac.in*

Bhurat, Swapnil S., *University of Petroleum & Energy Studies (UPES), Dehradun, India*

Chen, Wei-Hsin, *Department of Aeronautics and Astronautics, National Cheng Kung University, Tainan, 701, Taiwan; Research Centre for Smart Sustainable Circular Economy, Tunghai, 407, Taiwan; Department of Mechanical Engineering, National Chin-Yi University of Technology, Taichung, 411, Taiwan*

Dhanola, Anil, *Department of Mechanical Engineering, Faculty of Engineering and Technology, Guru Jambheshwar University of Science and Technology, Hisar, Haryana, India; E-mail: anildhanola3698@gmail.com*

Jain, Shubhansh, *Undergraduate, School of Mechanical Engineering, KIIT (DU), Bhubaneswar, Orrisa, India; E-mail: shubhanshjain28@gmail.com*

Liu, Zhongchuang, *Green Intelligence Environmental School, Yangtze Normal University, 16 Juxian Rd. Lidu, Fuling District of Chongqing, China*

Mahesh, G.S., *Geetha Shishu Shikshana Sangha Institute of Engineering and Technology for Women, Mysuru, India*

Nayak, Arunima, *Department of Chemistry, Graphic Era University, Dehradun, Uttarakhand, India*

Negi, Prateek, *Department of Mechanical Engineering, Graphic Era (Deemed to be) University, Dehradun, Uttarakhand, India*

Pal, Neha, *Department of Chemical Engineering, Malaviya National Institute of Technology Jaipur 302017, India*

Ranjit, P.S., *Department of Mechanical Engineering, Aditya Engineering College, Surampalem, India; E-mail: psranjit1234@gmail.com*

Raturi, Amit, *Department of Mechanical Engineering, Govind Ballabh Pant Institute of Engineering and Technology, Pauri Garhwal, Uttarakhand, India; E-mail: amitraturi4493@gmail.com*

Reddy, Sreenivasa, *Department of Mechanical Engineering, Aditya Engineering College, Surampalem, India*

Sharma, Abhishek, *Department of Mechanical Engineering, G L Bajaj Institute of Technology and Management, Greater Noida, Uttar Pradesh, India*

Singh, Yashvir, *Graphic Era Deemed to be University, Dehradun, Uttarakhand, India; Universiti Tun Hussein Onn Malaysia, Parit Raja, Batu Pahat, Johor, Malaysia; E-mail: yashvirsingh21@gmail.com*

Sisodia, Antriksh, *RMIT, University of Melbourne, Australia*

Thangadurai, Tavayogeshwary, *School of Chemical Engineering, Universiti Sains Malaysia, 14300 Nibong Tebal, Pulau Pinang, Malaysia; E-mail: chcttye@usm.my*

Tran, Khanh-Quang, *Department of Energy and Process Engineering, Norwegian University of Science and Technology, Kolbjørn Hejes vei 1B, Trondheim, Norway; E-mail: khanh-quang.tran@ntnu.no*

Tye, Ching Thian, *School of Chemical Engineering, Universiti Sains Malaysia, 14300 Nibong Tebal, Pulau Pinang, Malaysia*

Yadav, Avinash, *Department of Mechanical Engineering, Graphic Era (Deemed to be) University, Dehradun, Uttarakhand, India*

List of Abbreviations

EREVs	Extended range electric vehicles
GCI	Gasoline combustion ignition
H_2-ICE	Hydrogen fuel internal combustion engines
HEVs	Hybrid electric vehicle
HSVs	Hybrid solar vehicles
PFI	Ported fuel injection
PHEVs	Plug-in-Hybrid electric vehicles
RCCI	Reactivity controlled compression ignition

1

Current Scenario of Renewable Energy in India and Its Possibilities in the Future

Prateek Negi[1], Yashvir Singh[1,2], Avinash Yadav[1], Wei-Hsin Chen[3,4,5], Abhishek Sharma[6], and Antriksh Sisodia[7]

[1]Department of Mechanical Engineering, Graphic Era Deemed to be university, Dehrdaun, Uttarakhand, India
[2]Faculty of Mechanical and Manufacturing Engineering, Universiti Tun Hussein Onn Malaysia, Parit Raja, Batu Pahat, Malaysia
[3]Department of Aeronautics and Astronautics, National Cheng Kung University, Tainan, 701, Taiwan
[4]Research Centre for Smart Sustainable Circular Economy, Tunghai, 407, Taiwan
[5]Department of Mechanical Engineering, National Chin-Yi University of Technology, Taichung, 411, Taiwan
[6]Department of Mechanical Engineering, G L Bajaj Institute of Technology and Management, Greater Noida, Uttar Pradesh, India
[7]RMIT University, Melbourne, Australia
E-mail: yashvirsingh21@gmail.com

Abstract

Renewable energy resources and technologies have enormous potential to assist developing countries in addressing their longer-term energy challenges. While renewable energy sources such as wind, solar, geothermal, ocean, biomass, and fuel cell technologies may help India overcome its energy deficit, fossil fuels such as natural gas, coal, and oil will continue to be required. Renewable energy is one method to meet these criteria. Renewable energy currently accounts for a significant portion of India's main energy consumption. Renewable energy sources that are environmentally responsible are increasingly being utilized in India, resulting in more sustainable results for future generations. Over the last two and a half decades, India has made a concerted effort to explore a variety of renewable energy options for different sectors. The article examines India's renewable energy resources and makes new efforts to describe their availability, current state, major achievements, and future potential. Additionally, this research examines possible legislative solutions for eliminating barriers and expediting future renewable energy adoption.

Keywords: Datacenter design, Renewable energy, Challenges, Opportunities.

1.1 Introduction

According to the World energy forum, fossil-fuel resources will be exhausted in near future or over another decade. Over 75% of the primary energy used on the planet is sourced from fossil fuels, with over 57% of that energy consumed in the transportation sector, which is rapidly depleting and getting exhausted in near future [1]. The use of conventional energy and the depletion of natural resources are compelling policymakers and planners to explore alternative energy sources. Renewable energy is produced from resources that don't wear out over time. Renewable energy can help reduce carbon dioxide emissions by lowering the concentration of pollutants in the air while also improving air quality and setting the stage for future sustainability. SEIA promotes energy security and economic development while helping governments around the world to better protect themselves and their citizens. The modern-day equivalent of biomass is the use of chemical solar energy storage. Renewable energy sources account for 18% of global final energy consumption and comprise conventional biomass, big hydropower, and "new" renewables. Among the more popular alternative energy resources are micro-hydro, existing biomass, wind, solar, and geothermal energy, as well as biofuels. Traditional biomass, which is mostly used for cooking and heating, accounts for about 13% of energy and is not growing at a rapid rate. Hydropower represents 2% of total installed capacity in developing economies and is growing at a modest pace [2]. Renewable energy sources already account for about 2.4 percent of the global energy supply, and their share is growing rapidly in some developed and developing economies. Between 2002 and 2006, a variety of renewable energy technologies, including wind power, solar hot water, geothermal heating, and off-grid solar PV, saw year-over-year growth of around 15%–30% [3]. In 2008, the renewable energy industry saw considerable growth. Between 2005 and 2010, increased renewable capacity (excluding large hydroelectric dams) was mostly comprised of wind energy. Faced with a global investment of $120 billion in renewable energy in 2008, including new capacity (asset finance and projects) and biofuels refineries [4].

Indigenous renewable energy resources (RES) with zero emissions of air pollutants and greenhouse gases may be able to provide power with low emissions of greenhouse gases and air pollutants [5].

Several renewable energy technologies may utilize resources such as sunlight, wind, water, or biomass and transform them into commercial energy. As renewable energy resources can be competitive with conventional energy, their capacity to compete promises well for the future. In other words, this document was prepared to assist in helping people better understand renewable energy supply, existing supply, as well as the potential for renewable energy, overall, as well as all the government regulations, delivery, and outreach. India offers a wonderful image of using renewable energy resources on a world scale.

1.2 Renewable Energy

Despite its status as a developing country, India has enormous renewable energy potential. It is the world's fastest-growing economy, and its supply will demand an enormous quantity of energy [6]. The remedy to this energy issue is a larger energy demand; hence, renewable energy sources are the best alternative. India's MNRE (Ministry of New and Renewable Energy), is critical in encouraging renewable energy and reducing the effects of climate change. The following applications of renewable energy sources are being considered.

1.2.1 Biomass

Bioenergy is a sustainable energy source that also serves as an effective method of removing carbon dioxide from the environment. Biomass is commonly utilized for home cooking and heating in India. More than two-thirds of all energy is produced in the country. Wood, charcoal, and dried dung are the most common kinds of biomass. Rice husks, straws, coffee grinds, and dried manure are among other agricultural waste items. The power source is renewable, widely accessible, carbon-neutral, and may potentially help people in rural regions find work. Biomass can supply a stable kind of energy as well. Primary energy produced from biomass still accounts for

32% of the overall use, and around 70% of the population relies on it for their energy requirements. The emerging understanding of biomass energy's high propensity and importance in India has been recognized by the MNRE, and, as a result, has taken various initiatives to increase the deployment of more efficient biomass-based technologies across various sectors of the economy to guarantee the maximum possible return on investment. Under the new biomass power and cogeneration initiative, bagasse-based cogeneration and biomass electricity production have been integrated into sugar mills and combined cycle facilities. The combined cycle program by utilizing energy generated from biomass is being created to encourage the use of biomass resources for grid power production [7]. In producing biomass products, resources like bagasse, coconut husk, rice husk, crop leftovers, jute stalks, and de-oiled soya are all employed.

According to a recent study conducted on behalf of the Ministry of New and Renewable Energy (MNRE), India's current biomass stock is projected to be about 750 million metric tonnes each year. According to the findings of this research, there is an estimated surplus biomass capacity of about 230 million metric tonnes per year, which would generate approximately 28 GW of power production from agricultural waste. Concurrently, the country's 550 sugar mills, if run at the most cost-effective and technically optimal levels of cogeneration, could generate an additional 14 GW of power. Since the 1990s, a method for exploiting such resources for combined cycle power generation has been in development. In addition to the installation of 800 power projects and combined cycle plants by utilizing both bagasse and non-bagasse feedstocks, which supply a total of 10170 MW of energy to the grid, other biomass projects add 11250 MW to the system. The states that have led the charge in installing bagasse cogeneration facilities include Maharashtra, Karnataka, Uttar Pradesh, Tamil Nadu, and Andhra Pradesh, as well as Bhangar Thermal Power Station in Maharashtra, the potential is shown in Table 1.1. The leading states for biomass power projects are Chhattisgarh, Madhya Pradesh, Gujarat, Rajasthan, and Tamil Nadu.

Table 1.1 Installed biomass power in India (as of 30-6-2021) [8]

S. No.	Source	Potential (MW)
1.	Biomass	1836
2.	Bagasse based cogeneration	7562
3.	Non-bagasse based cogeneration	772
	Total	10170

Due to the installation of biomass power generation and bagasse-based cogeneration in sugar mills, many biomass power, and cogeneration initiatives have been launched. Although bagasse, a residual product in the sugarcane manufacturing process, is not particularly valuable, it is useful as a fuel for sugar mills due to its lower calorific value. Bagasse can be used to generate high-temperature and pressure steam, which can be used to manufacture sugar as well as generate electricity. The rise of rural economies increases the value of biomass. Because of the abundance of agricultural and plant waste, many people in rural regions have access to low-cost, ecologically beneficial energy. An empirical study found that expanding biomass power production is an efficient way to enhance energy security while simultaneously improving socioeconomic growth in east Asian nations [9]. Biomass is being employed as an energy source for a few objectives, including, to mention a few, facilitating payments, boosting global competitiveness, and boosting neglected rural economies. Purohit and Michaelowa [10] have performed research in which they assessed India's theoretical and attainable Clean Development Mechanism (CDM) potential for bagasse cogeneration. They determined that, if previous patterns of bagasse cogeneration are considered, bagasse cogeneration will not be possible for the country for another twenty years. Nonetheless, by adopting supporting CDM laws, it is possible to realize the greatest amount of potential in the shortest period.

Every year, India invests around $6 billion in biomass energy generation. The use of biomass in rural regions provides new job opportunities. India's Ministry of New and Renewable Energy (MNRE) has initiated nationwide programmes to increase biomass energy generation. The goal of this initiative is to encourage the adoption of new, creative technologies that will allow the country to make better use of its biomass resources. In India, the government has provided significant financial assistance to help overcome financial hurdles connected with biomass energy development. Regulatory incentives to boost biomass use, such as those created by NCDC (National Cooperative Development Corporation), IREDA (Indian Renewable Energy Development Agency), and government banks, are being efficiently implemented. The promotion of enhanced biomass-fuelled chulhas (stoves) seeks to contribute to the development of more technologically sophisticated, efficient, and environmentally-friendly models for biomass energy consumption. Numerous tax breaks for biomass energy generation, such as an accelerated depreciation rate of 80%, lower import and excise taxes, and others, have begun to entice investors.

When biomass is used as an energy source, it may bring several environmental benefits. Biomass energy is a powerful tool for reducing GHG emissions from both human and industrial sources. Massive amounts of methane are produced when biomass is burnt. Even though methane has a 20-fold shorter lifespan in the atmosphere than carbon dioxide, both pose serious risks. Methane can be collected and controlled from escaping into the atmosphere by collecting it from landfills, manure lagoons, and other sources. Methane collected can be utilized to create electricity or to power motor vehicles. Plants absorb carbon dioxide during photosynthesis and release it when biomass is burnt. Carbon sequestration may result in a higher carbon debt than carbon emissions from burning.

1.2.2 Biofuels

Gasification of biomass generates either wood gas or syngas, which may be converted into carbon-neutral methanol [11]. Currently, India produces about 750 million tonnes of non-edible biomass per year, which may be utilized for higher-value-added purposes and serve as a substitute for imported crude oil, coal, liquefied natural gas (LNG), urea fertilizer, and nuclear fuel, among other things. When biomass resources in India are used properly, it is anticipated that they would be able to replace all present fossil fuel usage [7]. Biomass will be critical in achieving India's energy independence and carbon neutrality [12].

Imported coal is used in fully operational pulverized coal-fired power plants. Because unpulverized biomass has a caking propensity, it cannot be used in pulverized coal mills. However, following torrefaction, biomass may be used instead of imported coal in pulverized coal mills. Because northern and southern regions may substitute torrefied biomass for imported coal when sufficient agricultural/crop residual biomass is available, the northwest and southern regions can become coal substitutes [13]. Renewable Purchase Certificates (RPCs) produced by biomass power plants may help the plants earn additional income by enabling the certificates to be sold (RPC) [14]. To reduce the carbon footprint associated with cement production, biomass that emits no carbon dioxide is being used [15]. Biogas, natural gas, or methane produced by cultivating the Methylococcus capsulatus bacteria near rural/consuming areas, along with minimal land and water footprint, may be used to feed cattle, fish, poultry, and pets, thus increasing animal production [16]. By using the CO2 gas generated by these machines, algal oil may be produced at a cheaper cost from algae or spirulina in countries such as India

[17]. Eleven second-generation ethanol plants are now under construction throughout the country to collect agricultural waste from farmers and convert it to bioethanol for India's three OMCs [18][19]. In 2018, India intends to build 5,000 large-scale commercial biogas plants, each having a daily capacity of 12.5 tonnes of bio-CNG, to generate 15 million tonnes of biogas and bio-CNG per year [20]. Biopropane is produced from non-edible vegetable oils, residual cooking oil, waste animal fats, and other organic resources [21].

1.2.3 Small Hydro

Hydropower facilities can be described in terms of size, and both big and small hydro plants are included. The capacity of small hydropower plants can vary anywhere from 10 MW to 50 MW in different places throughout the world. In India, small hydropower stations have capacities of 25 MW or less. Also, micro-hydro power has a capacity of 100 kW or less, mini-hydro power has a capacity of 101 kW to 2MW, and small hydropower has a capacity of 2MW or less (with a capacity of 2–25MW) [22]. Hydropower before 1989 was largely used as a resource by the State Electricity Boards, which was an important component of the Ministry of Power's mission. Under the jurisdiction of the Ministry of New and Renewable Energy (MNRE), small hydro projects with aggregate capacities of 3 MW or less came under the administration of the ministry in 1989, when their plant capacity was set at 3 MW or less. Although India has taken measures to increase hydropower development in hillier regions, such as through the establishment of a technical assistance project funded by the UNDP-GEF-AID, titled "Optimizing Development of Small Hydro Resources in Hilly Regions of India," over the past four years, the Indian Ministry of New and Renewable Energy has put considerable effort into this area. The MNRE was given 25 MW of electric capacity in November of 1999 [23].

According to their July 2016 study, "Alternate Hydro Energy Centre (AHEC) of IIT Roorkee Small Hydro Database," the country's estimated potential for electricity generation from small hydropower projects is about 211,35.37 MW. Over half of India's total landmass is in the mountainous states of Arunachal Pradesh, Himachal Pradesh, Jammu and Kashmir, and Uttarakhand. As a second option, Maharashtra, Chhattisgarh, Karnataka, and Kerala may all become independent states. Management's attention is attracted to various states because of interactions such as meetings, project monitoring, and assessing policy environments to attract private-sector investment money.

The Ministry has undertaken a variety of initiatives to promote predictable and dependable SHP development, as well as to enhance project quality and reliability. Investing in commercial SHP projects and assisting states with the development of small hydropower projects attracts more money to commercial SHP projects, which in turn promotes additional investment. Modern and upgraded hydroelectric generator models for mechanical and electrical output will be preferred to promote the establishment of micro-hydroelectric projects capable of generating up to 100 KW of energy for rural areas. Local organizations such as cooperative societies, registered NGOs, state nodal agencies, and village energy cooperatives are involved in these projects.

1.2.4 Solar Energy

Solar radiation and wind are the most abundant kind of persistent energy on Earth. People are generally mainly concerned with heating water, cooking food, drying wet things, purifying water,, generating power, and numerous other uses of thermal energy [24]. While light and heat are produced by the Sun, we also utilize solar energy through the photovoltaic process. The solar radiation that falls on Earth each year represents the total yearly energy supply, which is approximately 7,500 times more energy than the entire global annual primary energy consumption of 450 EJ. For many years, many people believed that renewable energy sources, including fossil fuels and nuclear, did not amount to more than three million Exajoules, although the total amount of solar radiation that reaches the Earth's surface each year exceeds three billion Exajoules. Fossil fuels provide 80 percent of the world's energy. The sun provides an average of 4–7 kWh of solar energy per square meter per day over a year, and approximately 250–300 days are sunny. West Rajasthan has the most yearly solar energy output, whereas the north-eastern part of the country has the lowest. Solar energy's recent impact on India's energy sector has been enormous. Solar lighting applications that are decentralized and distributed have provided thousands of people with reliable and renewable illumination in an economically and environmentally friendly manner. Rural women and girls who harvest and boil wood for fuel have less work, which results in an increase in community employment opportunities, an improved quality of life, and the ability to engage in economic activities. Additionally, solar energy has aided in the expansion of the grid-connected generation market in India. To meet the country's energy requirements, developing as a necessary component of the solution is also a viable option.

The country's solar potential, or the amount of solar energy that might be generated from unused land, is estimated to be 748 GW if 3% of that area is covered with solar photovoltaic modules. Solar Energy is a critical component of the country's National Climate Change Action Plan, with the National Solar Mission serving as one of the primary missions. On January 11, 2010, the National Solar Mission (NSM) was started. The National Solid Minerals Mission (NSM) is a large-scale effort in India, led by the federal government and including state governments, to achieve sustainable ecological development and energy security for the country. India will play a key role in the global effort to avert climate change via this endeavor. The Indian government has committed to becoming a world leader in solar energy, and the Mission's purpose is to spread the technology. Over 100 GW of grid-connected solar energy will be installed by 2022 to reach this aim. This is consistent with India's INDC goals of increasing non-fossil fuel energy capacity to 40% of total installed capacity by 2030 and cutting GDP emissions by 33–40% compared to 2005 levels [25].

India's primary objective is to produce as much solar energy as possible for domestic consumption. The government established a Renewable Purchase Obligation (RPO) goal as part of its policy efforts, which included a solar route. Guidelines for solar energy procurement through competitive bidding based on tariffs Photovoltaic solar standards, along with the installation of rooftop photovoltaic panels and preparedness for the rise of smart cities, a change to building by-laws requiring new construction or a higher FARR requirement for solar projects Long-term loans from multilateral financial organizations are a great method to fund solar bond projects that produce tax-exempt revenue. India has just overtaken Italy as the world's fifth-largest solar energy market. Since March 2014, when it was only 2.6 GW, solar energy capacity has grown by a factor of eleven. Grid parity has been achieved in India, and energy from the sun is now very cost-effective.

1.2.4.1 Grid-connected

Solar energy has emerged as a rapidly expanding sector in India in recent years. The government's aim is sustainable development, and as a critical component of addressing the nation's energy needs, emerging as an integral member of the solution to accomplish that goal is also a part of the solution.

The aim of constructing 100 GW of grid-connected solar energy by 2022 has been abandoned. One of the main initiatives of the Indian government is the Solar Park Scheme, which includes multiple other projects, such as

Table 1.2 Grid-connected solar power capacity in MW **[26]**

S. No.	Year	Total Capacity (MW)
1	Upto 2010	11.35
2	2010-11	35.93
3	2011-12	932.30
4	2012-13	1684.5
5	2013-14	2632
6	2014-15	3744
7	2015-16	6762.9
8	2016-17	12288.8
9	2017-18	21651.4
10	2018-19	28180.6

the VGF schemes, CPSU schemes, Defence schemes, Canal bank, canal top schemes, bundling schemes, and grid-connected rooftop schemes. Policies are being created to encourage the development of grid-connected solar energy systems. India has recently improved its position in solar energy deployment, rising from fifth to fourth place. Between 2014 and 2019, solar capacity almost doubled, from 2.6 GW to 28.18 GW, the year-wise description is shown in Table 1.2 Solar energy in India has become more affordable because of technological advances, making it competitive with other energy sources.

1.2.4.2 Off-grid solar PV program

The Ministry's initial initiative is to ensure that people have access to solar PV systems in places where grid power is not available or does not consistently work. This project aims to utilize all available solar electricity, whether it be used for lighting purposes, power generators, compressors, etc. Because it uses sunlight as input and energy as an output, solar photovoltaic (SPV) technology is very appealing. It looks to be the most promising of various decentralized power generation technologies. SPV systems are commonly utilised in off-grid regions for illumination and small-scale power generation. Currently, India is striving, with limited success, to deploy SPV systems. A slew of new efforts has been started to improve SPV system research and development, to boost the number of SPV applications. SPV cells are found in a wide range of devices, including solar lights and solar water pumping systems. In many cases, fossil fuels and other petroleum goods are more expensive.

One study revealed that solar home systems (SHSs) can reduce carbon dioxide emissions in India. According to the findings, they are functioning

Table 1.3 Off-grid installation status [27]

S. No.	System	Units
1	Solar lamps	6517180
2	Solar pumps	237120
3	Solar streetlights	671832
4	Solar Home lighting systems	1715639

at a level below their capabilities. Due to the significant capital expenditures, SHS is at a great disadvantage. However, SHS installations are well-suited to supporting emission reduction because of their good suitability, but the role they may play in the development of rural regions may be maximized by implementing aggressive and targeted policy interventions. Purohit and Michaelowa's following study [17] revealed that government subsidies are necessary to maximize SPV pumps' CDM potential.

A 2000 MWp objective was established for off-grid solar PV applications under the National Solar Mission. During the first phase, which ran from 2010 to 2013, the aim was 200 MWp. This was fulfilled, as well as exceeded, by the installation of 253 MWp, and during the second phase, which started in 2013 and will continue through 2017, the target is 500 MWp. A total of 118 MW has been designated as the maximum cap for decentralized and off-grid solar PV installations under Phase III of the Off-grid and Decentralized Solar PV Applications Scheme, excluding pumps and solar home lights acquired through the Ministry of Power's "Saubhagya" program as shown in Table 1.3.

The country's solar off-grid program is dependent on solar pumps, which can dependably water crops in remote locations. In general, solar water pumps are well-suited for small farms. To finish the installation of the new solar-powered pump, the previous diesel pump was replaced with a new one that is now powered by the sun. MNRE began to implement the project in 1992 with the implementation of its program. During the period from 1992 to 2014, the U.S. put over 11,600 solar pumps in place. One million solar pumps with a total investment of 400 billion rupees were built in India in the 2014-15 fiscal year for irrigation and drinking water applications. The implementation of the solar pump project has brought down the average cost of solar pumps by around 30 percent. Until March 31, 2017, portable solar pumps could be used with off-grid and decentralized solar PV systems. Pradhan Mantri Kisan Urja Suraksha evam Utthan Mahabhiyan, or PM KUSUM, is the name of a new program by the Indian government to give financial assistance to farmers to construct new, freestanding solar pumps in off-grid locations, and connect their old irrigation water pump to the power system. This would ensure farmers have a consistent supply of irrigation, thus improving their income and overall economic position. The government has also initiated a program to distribute Seventy million Solar-based Reading Lightbulbs to rural schools. These lamps can provide excellent, low-cost light for the students, while also helping the environment. At the time of the 2011 census, the following states lacked electricity for more than 50% of their population: Assam, Bihar, Jharkhand, Odisha, and Uttar Pradesh, which have all implemented a Scheme. The program benefits more than half of kerosene-using households. Students are expected to pay just Rs.100 towards the lamp's total cost of roughly Rs.450, with the remainder subsidized by the government. The Ministry of New and Renewable Energy (MNRE) is subsidizing several states as they construct solar LED streetlights in rural, semi-urban, and metropolitan regions under the Atal Jyoti Yojana (AJAY), state-wise schemes are shown in Table 1.4. Solar LED streetlights can offer sufficient light for those living in remote areas, allowing them to live in safety and security.

1.2.5 Wind Energy

Wind energy is at the cutting edge of renewable energy generation. Despite its lengthy coastline, India has the adequate wind to generate energy. According to the Renewable Energy Foundation, India has the greatest potential for wind energy of any renewable energy source. India is the world's fifth-greatest wind energy producer, behind the United States, Germany, Spain, and China. India's indigenous wind energy industry is economically and socially more significant than the country's wind energy sector. The wind energy sector now has 10,000 MW of yearly capacity due to the industry's growth. Currently, the country ranks fourth in the world in terms of wind capacity deployed. Wind power produced 60.149 billion units in the year ending March 31, 2021. By giving different fiscal and financial advantages, the government is making the private sector wealthy and incentivizing wind power development across the nation by offering accelerated depreciation benefits. Another option for projects completed before March 31, 2017, is the Generation Based Incentive (GBI) scheme.

The Government supports projects funded by private sector investment via Accelerated Depreciation and concessional customs tax exemptions on certain components of wind turbines, which offer fiscal and financial benefits to the whole country. Keeping this in mind, it's worth noting that the GBI program is only available for wind projects built before March 31st, 2017.

Table 1.4 Details of individual schemes (up till 2019) [27]

States	Lanterns	Home Lights	Streetlights	Pumps	Stand Alone
Andhra Pradesh	77803	22972	8992	34045	3815.595
Arunachal Pradesh	18551	35065	5008	22	963.2
Assam	498361	46879	9547	45	1605
Bihar	1258294	12303	29858	2107	6770
Chhattisgarh	3311	42232	2042	61970	31249.9
Delhi	4807	Not available	301	90	1269
Goa	1093	393	707	15	33
Gujarat	31603	9253	2004	11522	13577
Haryana	93853	56727	34625	1293	2321
Himachal Pradesh	33909	22592	78000	6	1906
Jammu & Kashmir	51224	144316	14156	39	8130
Jharkhand	747295	9450	10301	3857	3770
Karnataka	7781	52638	2694	6343	7754
Kerala	54367	41912	1735	818	15825
Madhya Pradesh	529101	7920	10833	17813	3654
Maharashtra	239297	3497	10420	4315	3858
Manipur	9058	24583	11205	40	1581
Meghalaya	40750	14874	5800	19	2004
Mizoram	10512	12060	5325	37	2956
Nagaland	6766	1045	6235	3	1506
Orissa	99843	5274	14567	9327	568
Punjab	17495	8626	42758	3857	2066
Rajasthan	225851	187968	6852	48175	30349
Sikkim	23300	15059	504	0	850
Tamil Nadu	16818	290376	39419	4984	12753
Telangana	0	0	1103	424	7450
Tripura	64282	32723	1199	151	867
Uttar Pradesh	1336733	235909	258863	20465	10638
Uttarakhand	163386	91595	22119	26	3145
West Bengal	17662	145332	8726	653	1730
Andaman & Nicobar	6296	468	390	5	167
Chandigarh	1675	275	898	12	730
Lakshadweep	5289	600	2465	0	2190
Puducherry	1637	25	417	21	121
Others	125797	24047	9150	609	23885
NABARD	0	116226	0	4012	0
Total	5823800	1715214	659218	237120	212054

Table 1.5 The capacity of wind in different states of India.

S.No.	State	Wind Potential at 100 m (GW)	Wind Potential at 120 m (GW)
1	Gujarat	84.43	142.56
2	Rajasthan	18.77	127.75
3	Maharashtra	45.39	98.21
4	Tamil Nadu	33.79	68.75
5	Madhya Pradesh	10.48	15.40
6	Karnataka	55.85	124.15
7	Andhra Pradesh	44.22	74.90
	Total 7 windy states	292.97	651.72
8	Others	9.28	43.78
	Total	302.25	695.50

Additionally, the following steps have been implemented to aid in the country's wind energy deployment:

a) Along with wind resource appraisal and site selection, technical support, including wind resource assessment, is provided via the National Institute of Wind Energy in Chennai.
b) Costs and losses associated with the interstate transmission have been removed for wind and solar projects scheduled to begin construction by March 2022, enabling interstate commerce in wind energy.
c) Wind energy procurement framework with transparent and standardized process based on tariff-based competitive bidding for grid-connected wind energy. These rules let Distribution Licensees get cheap wind energy.

Wind energy is difficult to locate due to its intermittent and local nature, and therefore a comprehensive wind resource evaluation is needed for the site selection process. Table 1.5 shows the capacity of wind in India. Wind potential maps at five heights have been prepared by the National Institute of Wind Energy (NIWE): 50 meters, 80 meters, 100 meters, 120 meters, and 150 metres above ground level. A recent study of the country's wind energy resources found that they total over 300 GW at ground level, over 700 GW at 100 m, and 695.5 GW at 120 m. The bulk of this under utilized resource is in the seven states indicated below [28]:

1.2.6 Waste to Energy

Accompanying economic development comes the creation of increasing quantities of garbage, posing increasing dangers to the environment. To aid in the generation of large amounts of decentralized energy and to aid in the prevention of waste, new methods have been used in recent years [29].

The Ministry of Renewable Resources is sponsoring a variety of emerging solutions to enhance the recovery of energy from agricultural, industrial, and urban waste, including municipal solid waste, agricultural leftovers, and industrial/STP waste and effluent.

Numerous different types of trash may be generated during our daily activities or as a consequence of industrial activity. This category includes organic waste, e-waste, hazardous waste, and inert rubbish. Organic waste is trash that degrades or is broken down by microorganisms over time. All organic wastes are essentially carbon-based compounds. Despite this, they may differ significantly in terms of appearance and rate of disintegration. A significant portion of industrial, urban, and agricultural waste is comprised of organic compounds, which may be used to generate energy [30]. Non-biodegradable and biodegradable organic waste, which accounts for the majority of trash's organic components, may be further classified as non-degradable and degradable.

(a) Biodegradable trash is comprised of organic materials that can be rapidly degraded into nutrients for absorption by natural microbiological microorganisms. Biodegradable organic materials include inorganic materials (organic waste from poultry farms, cattle slaughterhouses, dairy, sugar, distillery, paper, oil extraction plants, starch processing plants, and leather sectors).

(b) Non-biodegradable organic compounds are those that are not biodegradable or have a very slow biodegradation rate. Plants and other cellulose-based materials (e.g. cardboard, wooden furniture, bags, wrappings, clothes, agricultural dry waste, bagasse, rice husk)

The various technologies available to convert waste into energy to obtain electricity and biogas are described below [31]:

(i) *Biomethanation* is the anaerobic digestion of organic matter into biogas. Anaerobic Digestion (AD) produces biogas, which is mainly methane (around 60%) and carbon dioxide (approximately 40 percent). Biomethanation has two benefits. Biogas is the primary product, followed by manure. This system is ideal for processing segregated organic wet waste, such as kitchen waste, institutional waste, hotel waste, and slaughterhouse waste. Biomethanation biogas may be used to generate energy or heat for industries like thermal processing and cooking. Biogas may be processed to remove CO2 and other contaminants before being used to produce BioCNG. It may be injected into the national gas system for use as vehicle fuel. Using Biomethanation, 20–25 kilograms of cow dung may produce 1 cubic meter of biogas, which can provide 2 units of energy or 0.4 kilograms of BioCNG.

(ii) *Incineration:* Complete combustion of garbage (refuse produced fuel) and heat recovery are accomplished via incineration technology, which generates steam that is utilized to power steam turbines that generate electricity. As a result, a sophisticated air pollution control system is needed to purify the flue gases produced by the boilers. The combustion of solid waste produces residue that may be disposed of appropriately in a landfill. This method has been extensively utilized in India for a variety of waste kinds, with remarkable results.

(iii) *Gasification* is a process that melts biomass to extreme heat (500–1800 0C) and requires a minimal amount of oxygen to breakdown it into synthetic gas (H2). Syngas is made from agro-residues, segregated MSW, and RDF pellets. Another way to use this gas is through thermal or electrical generating. The primary goal of garbage gasification is to reduce the cost and harmful impact on the environment.

(iv) *The pyrolysis* process, which uses heat to separate compounds that do not contain oxygen, results in the production of a combustible gas mixture that contains mostly methane, complex hydrocarbons, hydrogen, and carbon monoxide. a gaseous mixture; a liquid (bio-oil/tar); or a solid waste product (carbon black). The gas produced by either of these processes may be used in boilers to generate heat, or it can be cleaned up and used in combustion turbine generators to generate electricity. The main goal of the pyrolysis of trash is to reduce emissions while simultaneously increasing the quantity of resources that can be recovered.

The following is a summary of the sector-wise potential for India's urban and industrial sectors, which are primarily concerned with energy. In India, the projected potential for energy production from urban and industrial organic waste is about 5690 MW, as shown in Table 1.6.

1.2.7 Geothermal Energy

Geothermal energy is produced when heat is stored in the Earth's crust or extracted from the Earth's subsurface. The Earth is one of the hottest places in the universe due to the enormous amount of thermal energy generated and stored in the Earth's core, mantle, and crust, as well as the enormous amount

Table 1.6 Details of waste and it's potential for energy.

S. No.	Sectors	Energy potential–MW
1	Municipal Solid Waste	**1247**
2	Municipal Liquid waste	**375**
3	Rag waste (fluid)	**254**
4	Meat processing and storage (fluid)	**182**
5	Meat processing and storage (solid)	**13**
6	Storing and processing seafood (liquid waste)	**17**
7	Processes Associated with Vegetables (solid waste)	**3**
8	Vegetables in their natural state (solid waste)	**579**
9	Preparation of Fruit (solid waste)	**8**
10	Fruit in its natural state (solid waste)	**203**
11	Palm Oil (solid waste)	**2**
12	Processing of Dairy Products (liquid waste)	**24**
13	Maize Starch (liquid waste)	**47**
14	Tapioca Starch (liquid waste)	**36**
15	Tapioca Starch (solid waste)	**15**
16	Sugar (liquid waste)	**49**
17	Sugar press mud (solid waste)	**200**
18	Distillery (liquid waste)	**781**
19	Slaughterhouse (solid waste)	**48**
20	Slaughterhouse (liquid waste)	**263**
21	Cattle farm (solid waste)	**862**
22	Poultry (solid waste)	**462**
23	Chicory (solid waste)	**1**
24	Tanneries (liquid waste)	**9**
25	Tanneries (solid waste)	**10**
Total (MWeq)		5690

of thermal energy produced and stored in the Earth's atmosphere. The world and India currently receive 10,000 MW of geothermal energy, even though nature only supplies a finite amount of such energy. According to research conducted by the Geological Survey of India, around 340 hot springs have been discovered throughout India. These engines are distributed across the country in seven regions and are fuelled by geothermal energy. Although these nations are located west of the Himalayas, they span the entire 1500-kilometer range. The long-term plan is to triple the resource's capacity during the following four years [32].

India is trying to identify and exploit the country's geothermal energy potential. Geothermal resources may be used effectively for a wide range of applications, including power generation, space heating, agriculture, and other activities. The National Geophysical Research Institute (NGRI) in Hyderabad, India, has traced and evaluated geo-electric structures in Tatapani in Chhattisgarh, Surajkund in Jharkhand, Badrinath-Tapovan in Uttarakhand, Satluj-Beas, and Parvati Valleys in Himachal Pradesh, and Puga in Jammu and Kashmir's Ladakh region [33].

1.3 Future of Renewable Energy in India

The Indian government is making high goals and ambitions for expanding the amount of renewable energy (RE) in the country's energy reserves [34]. India intends to construct 175 GW [35] of renewable energy facilities by 2022, with a total capacity of 450 GW by 2030 [36]. To put things in this context, the total installed energy capacity of India by the finish of 2020 was 379 GW, with renewable energy accounting for 93 GW (25 percent). Until recently, large grid-scale solar has been the primary focus of the government's renewable energy initiative. Even if India's ambitious renewable energy (RE) targets are fulfilled, additional distributed renewable energy (DRE) projects would be needed. Although modest in size, such DRE efforts may have more growth potential if a more favorable legal and regulatory environment is created. Furthermore, public-sector offtake procurement efforts are hampered by long lead times and execution challenges.

This plan calls for about 175 GW of renewable energy capacity, of which 100 GW is solar and 40 GW is residential rooftop solar (RTS) and off-grid solar (OGS). Many revenue streams are available for rooftop and off-grid games. The government has already implemented certain structures that cover a broad range of downstream OGS applications, including agricultural pumps, cold storage, and home systems, as well as a broad range of downstream DRE applications, including energy storage, electric vehicle charging, and non-farm productive applications in rural areas. DRE, as a result, should play a critical part in India's efforts to attain energy sustainability during the next ten years. Annual investments will need to more than double from USD 2 billion in 2019 to USD 18 billion by 2024 to meet these objectives, Figure 1.1 shows the DRE's annual financing market report. To support that degree of investment growth, the current DRE policy framework would have to be significantly changed.

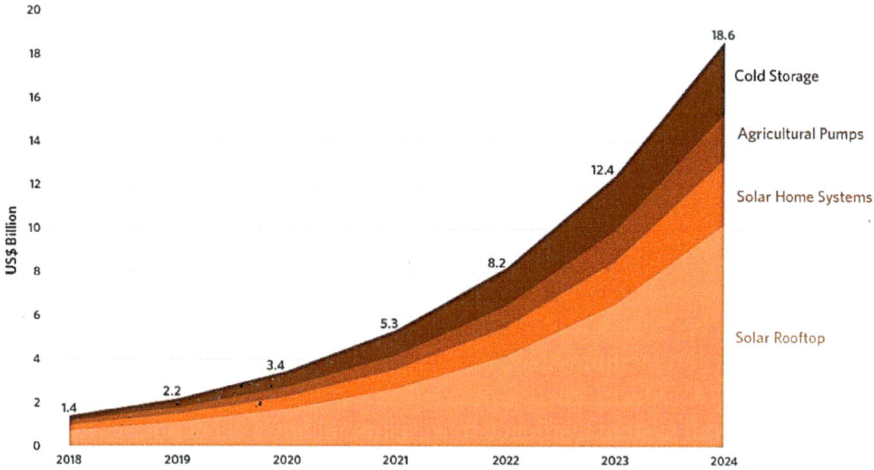

Figure 1.1 DRE's annual financing market report.

However, governments backed by central financing first created a favorable environment for DREs. Many of these incentives, however, have been progressively phased off in recent years. Because state energy distribution companies (DISCOMs) have had their RTS subsidy rollbacks delayed, they are beginning to view RTS as a risk to their finances, raising their costs, and a longer-term barrier to disintermediation. While all stakeholders are feeling the effects of the COVID-19 pandemic, it has also had an impact on their business strategies and forecasts of future financial flows. To ensure that DISCOMs continue to play an important role and generate revenue, the GoI must rethink its approach to DRE and establish an active and varied private market for different downstream applications.

1.4 Policy Gaps and Opportunities

To achieve India's long-term energy targets, the government needs to refocus its policies and pursue a healthy private DRE market.

A clear and predictable legal framework in which all stakeholders are rewarded is required to help push more public and private investment into the DRE industry. Specifically, some of the extra instances generated in this research are as follows:

Rooftop Solar: The government will empower DISCOMs to collect transaction fees in addition to monthly Operation & Maintenance costs by implementing a demand aggregation model as part of India's Phase II grid-connected RTS scheme. By remaining connected to this plan, the DISCOMs would be able to maintain relevance and avoid disintermediation.

Distributed Storage: The Phase II RTS system should include a strategy for distributed energy storage. Rather than supporting a capital-subsidy-based model, the government should promote an environment that is more favorable to operational models, such as DISCOMs.

Smart Energy Management: Incentives for IoT-based energy efficiency retrofits that can be connected to existing home circuits would help consumers and small and medium-sized businesses reduce their energy use. One advantage is that it reduces energy prices and carbon footprint, but it may also improve overall system resilience by assisting the grid is becoming more robust. For example, DISCOMs may wish to go even faster in the direction of Time-of-Day pricing, which is a demand-side management approach.

Electric Vehicle Charging Infrastructure: Although electric vehicle charging services can be considered a public good, the government should support it. It is critical to employ decentralized tactics and to encourage the adoption of franchise-based structures monitored by distribution corporations. Allowing businesses that generate additional solar energy through RTS to build retail charging stations would also be an excellent move.

Solar Agricultural Pumps: The government's KUSUM plan presently employs a centralized tender process. Policymakers should carefully consider enabling state DISCOMs to work with private installers on a neighborhood-by-neighborhood basis. Commercial users and solar pump installations, as well as local farmer co-operatives, may enter business partnerships. In other words, if the DISCOM includes a solar pump with the installation, it may collect excess power generated by the pump and pool it into a single point of injection into the grid, from which it may then pay farmer cooperatives the energy purchase prices, net of any service costs.

Solar Cold Storage: The GoI offers a 30% discount on solar cold storage systems as part of the rural livelihood subsidy program. However, since the agricultural supply chain depends significantly on cold storage, it is essential to create a unique solar cold storage program for lowering capital costs.

Financing Gaps and Opportunities: Both the RTS and OGS industries are modest and have not seen significant growth as a result of their reliance on charity or subsidized private support. Currently, the business is attracting just a small amount of private commercial investment. To further boost DRE market growth, the Indian government, along with the public, private, and development/charitable sectors, worked together to promote greater financial sector intervention and prudent public investment.

DRE and its downstream applications provide possibilities that go beyond fulfilling India's climate and renewable energy objectives; they are also great investments with high yields. Many smaller RTS and OGS game production companies have improved their business strategies in recent years and are now seeking for growth funding. Despite this, smaller RTS and OGS developers are at a disadvantage since many lenders continue to put limitations on them when it comes to utilizing credit ratings. Their prospects of obtaining the necessary financing are further hampered by a lack of expertise. While there is a need for further rounds of early-stage financing, technical assistance, and strategic advice for such companies, it also grows because of their ever-increasing need for such resources.

Small DRE developers are also unable to get finance because of a lack of project planning and focused transaction assistance. Accelerated government and philanthropic support are required to help address these gaps, allowing the DRE sector to grow quickly.

Despite growing DRE investment capital in the private sector, impact investors such as family offices, high net-worth individuals, and businesses have only slightly expanded their DRE holdings over the past decade. DRE is seen as less attractive to investors with a significant quantity of cash to invest due to the requirement for money in traditional business methods. Mixed financial products may help to bridge this gap.

Regardless of their objective, commercial financial investors and impact investors often favor mature-stage companies and projects. Smaller and growing companies in specialized areas with the potential to expand over the next few years may be rewarded and attract greater investment attention.

DRE implementation provides a once-in-a-lifetime opportunity to meet India's climate and renewable energy goals while simultaneously giving significant financial benefits to investors. As a result, because the DRE reduces the likelihood of further dependency on oil imports, India can be more confident in the long run and produce more economic growth and jobs. If you support government spending programs that are more focused and risk-averse while still recycling money efficiently, developing a combination of policy advocacy, information distribution, and catalytic finance may allow those programs to survive longer and generate greater results.

1.5 Conclusion

Power and prosperity, as well as environmental stewardship, all have an impact on global energy policy. The recent increase in the price of crude oil necessitates a greater effort to promote and develop alternate energy sources. We must support renewable energy sources to address issues such as economic growth, energy security, and environmental degradation. Members of the NAPCC advocate for renewable energy projects that incorporate a variety of energy sources. Numerous different metrics have been highlighted, including assisting in the promotion of renewable energy development and making it easier for renewable energy technologies to join the market and new renewable energy technologies to be developed. Further coal-fired power plant capacity would be unnecessary because the capacity is now under development and expected to be completed between 2017 and 2022. As fossil fuels dwindle and pollution levels rise, renewable energy sources that are non-renewable, clean, and green are the best option for meeting energy demand in today's world. Increased use of biofuels, mostly as a blending component with diesel and gasoline, is anticipated as the transportation sector's demand for those fuels grows. When economic, environmental, and societal factors are taken into account, renewable energy is far superior to conventional energy sources. Renewable energy's estimated share of total producing capacity is expected to grow in the future.

References

[1] International Energy Agency IEA. Key world energy statistics. Available at: http://www.iea.org/Textbase/nppdf/free/2006/Key2006.pdf [Accessed: 18/08/2021].

[2] World Energy Outlook. International energy agency; 2008. http://www.worldenergyoutlook.

[3] org/2008.asp.

[4] REN21, Renewables 2007 global status report. http://www.ren21.net/pdf/.

[5] REN21, Renewables 2009 global status report. http://www.ren21.com.

[6] Varuna SK, Singal. Review of augmentation of energy needs using renewable energy sources in India. Renewable and Sustainable Energy Reviews 2007; 11:1607–15.

[7] Kumar A, Kumar K, Kaushik N, Sharma S, Mishra S. Renewable energy in India: current status and future potentials. Renewable and Sustainable Energy Reviews 2010;14(8):2434–42.

[8] Yee, Amy (8 October 2013). "India Increases Effort to Harness Biomass Energy (Published 2013)". The New York Times. ISSN 0362-4331. Accessed July 2021

[9] Ministry of New and Renewable Energy. Available at: https://mnre.gov.in/bio-energy/current-status [Accessed: 17/8/2021].

[10] Bazmi AA, Zahedi G, Hashim H. Progress and challenges in utilization of palm oil biomass as fuel for decentralized electricity generation. Renewable and Sustainable Energy Reviews 2011;15(1):574–83.

[11] Purohit P, Michaelowa A. CDM potential of bagasse cogeneration in India. Energy Policy 2007;35(10):4779–98.

[12] "Renewable Methanol" (PDF). Retrieved 19 May 2021

[13] Future of biomass economy in carbon neutral India". Retrieved 19 December 2020.

[14] "CEA has written to all States to use 5-10% of biomass pellets with coal for power generation in thermal power plants". Business Standard India. 8 February 2018. Retrieved 22 February 2020. -85

[15] "Renewable purchase obligations enforcement is not our remit: Power regulator". Retrieved 6 April 2017. -87

[16] "New IEA Report: Renewable Energy for Industry". Retrieved 21 June 2018.-89

[17] "360-degree plan to convert cattle dung into energy". Retrieved 22 February 2018. -91

[18] ExxonMobil Announces Breakthrough In Renewable Energy". Retrieved 20 June 2017. -96

[19] "India can replace Rs 1 lakh crore worth of hydrocarbon imports by bio-fuels: Pradhan". Retrieved 8 July 2017.-97

[20] "Bio-fuel business to pick up in India: Atul Mulay, Praj Industries". Retrieved 30 March 2018.-98

[21] "Clean push: Why compressed biogas has an edge over CNG". Retrieved 11 February 2021. -100

[22] "47 lakh kg used cooking oil collected since Aug; 70% converted into bio-diesel". Retrieved 29 December 2019. -101

[23] Ministry of New and Renewable Energy. Available at: https://mnre.gov.in/small-hydro/current-status [Accessed: 17/8/2021].

[24] Nautiyal H, Singal SK, Varun, Sharma A. Small hydropower for sustainable energy generation in India. Renewable and Sustainable Energy Reviews 2011; 15:2021–7.

[25] Urja Akshay. Newsletter of the Ministry of New and Renewable Energy. Government of India; December 2008. http://mnes.nic.in/akshayurja/novdec-2008-e.pdf.

[26] Ministry of New and Renewable Energy. Available at: https://mnre.gov.in/solar/current-status/ [Accessed: 5/5/2021].

[27] Ministry of New and Renewable Energy. Available at: https://mnre.gov.in/solar/solar-ongrid [Accessed: 5/5/2021].

[28] Ministry of New and Renewable Energy. Available at: https://mnre.gov.in/solar/solar-offgrid [Accessed: 5/5/2021].

[29] Annual report 2020–2021, MNRE, Govt of India.

[30] Emmanual, William. "Energy Alternatives India". EAI. Retrieved 5 March 2012.

[31] Electricity from sewage in India, www.clarke-energy.com, retrieved 15 August 2014

[32] Emmanual, William. "Energy Alternatives India". Energy Alternatives India. Retrieved 5 March 2012.

[33] Ashwani Kumar, Kapil Kumar, Naresh Kaushik, Satyawati Sharma, Saroj Mishra, Renewable energy in India: Current status and future potentials, Renewable and Sustainable Energy Reviews, Volume 14, Issue 8, 2010, Pages 2434-2442, ISSN 1364-0321, https://doi.org/10.1016/j.rser.2010.04.003

[34] Himanshu Nautiyal, Varun, Progress in renewable energy under clean development mechanism in India, Renewable and Sustainable Energy Reviews, Volume 16, Issue 5, 2012, Pages 2913-2919, ISSN 1364-0321, https://doi.org/10.1016/j.rser.2012.02.008.

[35] Maithani PC. Renewable energy policy framework of India. India: Narosa Publication Delhi; 2008. p. 41–54.

[36] Saluja, Nishtha; Singh, Sarita (5 June 2018). "renewable energy target now 227 GW, will need $50 billion more in investments, MNRE". The Economic Times.

[37] Sep 23, PTI |Updated; 2019; Ist, 22:04. "PM Modi vows to more than double India's non-fossil fuel target to 450GW |India News - Times of India". The Times of India. Retrieved 23 September 2019.

2

Application of Green Nanomaterials for Sustainable Energy Systems: A Review of the Current Status

Brij Bhushan and Arunima Nayak

Department of Chemistry, Graphic Era University, Dehradun, Uttarakhand, India
E-mail: brij.chem@geu.ac.in

Abstract

The world is plagued with problems related to environmental pollution, ever-increasing demand for energy in the face of rapidly depleting natural resources like fossil fuel reserves, clean drinking water, and air. In order to address the gap between the increasing demand and supply of energy, nanotechnological application via synthesis of tailored nanoparticles has ensured innovations and advancements in various sectors of energy ranging from generation to conversion to transmission to storage. The unique properties of nanomaterials like excellent electrical and thermal conductivity, large surface area, and enhanced stability have expanded their application in the energy sector. The success achieved in research and development activities with respect to the synthesis of size-controlled nanomaterials, expanding energy-related application of such nanomaterials, and increased activities to make the energy activities cost-effective are the key factors that are driving the future market growth of energy-related nanomaterials. But despite the great success attained, various factors have been identified that have restrained the demand and market growth of energy-related nanomaterials. Green nanotechnology and the requirement for sustainability in diverse energy areas are the need of the hour. The outline of the chapter is thus to initially highlight the increased demand for nanotechnology vis-à-vis the energy domain along with focussing on the different areas of energy in which nanotechnology has made a major impact. This is followed by a discussion on the toxicity effects due to the conventional synthesis methods employed, due to the fabrication of nanomaterials onto different energy devices, and lastly, the fate of such nanomaterials during the consumerism of such devices. The need for sustainability in the energy sector and the importance of green nanotechnology are thus highlighted in the chapter. Finally, a critical discussion on different strategies that have been adopted to increase the sustainability of nanoparticle synthesis as well as on the applications is presented. The chapter is concluded with an eye-opener on certain critical issues like the lack of studies addressing the occupational health hazards associated with the use and exposure of energy-related green nanomaterials and safety aspects which otherwise can boost the wider acceptance and adoption of green nanotechnology in the energy sector.

Keywords: Green nanotechnology, Green Chemistry, Sustainability, Energy application, Nanoparticles.

2.1 Introduction

The energy sector has attracted enormous research and development activities mainly because of its contradiction to the area of sustainable development. This is because of the unfavorable energy demand-supply ratio, undue pressure on fossil fuel-based natural resources, and increasing environmental pollution. With increasing consumerism and lifestyle changes, there has been excessive demand for production, conversion, transmission, and energy storage. The International Energy Agency has forecast that the demand for energy is expected to rise by approximately 50 percent by 2030 [1]. Currently, fossil fuels are being fulfilled over 80 percent of the energy demand. But fossil fuel resources are dwindling and are becoming limited. Also, because of rising awareness among people regarding environmental pollution, global warming, and climate change due to indiscriminate use of fossil fuel energy resources, research has focussed on the search and use of non-renewable sources of energy. Sustainable development in the energy sector is also increasingly drawing the global population's attention, mainly because the current energy pathway is diverting away from the ideal sustainable concept. Both energy security and sustainable development hinge on economic growth, ecological balance, and environmental sustainability. Thus, addressing the challenges imposed by society, accelerating the economy, and ensuring sustainability in the energy sector, the focus of research is on renewable energy sources, energy security, energy pricing, energy policy, renewable energy applications, and smart grid technologies. The current enabling technologies are advanced materials, nanotechnology, micro and nanoelectronics, photonics, advanced manufacturing system, and industrial biotechnology. Nanotechnology has ensured more efficient, cost-effective, and environment-friendly production, transmission, and storage.

2.2 Use of Nanotechnology for Improved Energy Efficiency

Nanotechnology is the branch of technology that deals with developing nanoscale dimensions of materials and their application [2]. Nanotechnology in the energy sector involves the incorporation of nanomaterials for the development of new and improved energy technologies that can enhance the betterment of life across the globe. Some of the involvement of nanotechnology in the energy sector is in nanofabrication, energy production, transmission, storage, etc. [3]. With the advent of nanotechnology, the last two decades have seen the development of various nanomaterials and their application in energy devices for generation (photovoltaics, wind and geothermal, hydro/tidal power plants), conversion (gas turbines, fuel cells, steam power plants, etc.), transmission (high voltage power transmission, superconductors, etc.) and storage of electricity (lithium-ion batteries and supercapacitors). Table 2.1 summarizes and highlights the different areas of energy in which nanotechnology has brought about a definite impact.

Nanotechnology has improved the design and efficiency of both conventional and non-conventional energy sources [1]. For e.g. carbon nanotube, nanocomposites have been used to enhance the performance of rotor blades of wind and tidal power plants and bring about anti-corrosiveness and wear resistance for bearings, gear boxes, and other drilling equipment. The superior photo-oxidation capability of semiconducting nanomaterials used in solar hydrogen generation has been cited due to their higher surface area, higher optical absorption, shorter charge migration, and higher solubility than their bulk counterparts. Low solar energy for electricity generation and high fabrication costs are some of the limitations of photovoltaics that are causing limited commercial applications. The use of carbon nanotubes, fullerenes, and carbon dots are some of the nanomaterials whose usage has improved the functioning of photovoltaics [3]. Anti-reflective nano-coatings on silicon solar cells have enhanced light yields and hence resulted in maximizing the efficiency of conventional solar cells [1].

Table 2.1 Enhanced performance in the use of nanomaterials in different areas of energy application

<table>
<tr><td>Non-conventional energy source</td><td colspan="2">Solar energy/photovoltaics</td><td colspan="3">Wind/tidal energy</td><td colspan="2">Geothermal energy</td></tr>
<tr><td></td><td colspan="2">Carbon nanotubes, fullerenes, quantum dots, and graphene for enhanced efficiency in photovoltaics
Anti-reflective Nano-coatings for enhanced solar light capture
Nano-TiO_2 for enhanced conversion efficiency in Gratzel solar cell</td><td colspan="3">Carbon nanotubes, nano-silica, graphene, and nano-thermal coatings for enhancing mechanical properties and anti-corrosion in rotor blades in wind power station</td><td colspan="2">Nano-coatings, nanocomposites for enhanced corrosion resistance in drilling equipment</td></tr>
<tr><td>Energy conversion</td><td colspan="2">Steam power plants</td><td colspan="3">Coal fired power plants</td><td colspan="2">Fuel cells</td></tr>
<tr><td></td><td colspan="2">Use of carbon nanotube, ceramic or inter-metallic coated turbine blades for high resistance to heat and wear</td><td colspan="3">The use of nano-membranes for the capture of CO_2 has enabled environmental friendliness</td><td colspan="2">The use of nano-structured electrodes or membranes has boosted energy yield</td></tr>
<tr><td>Energy distribution</td><td colspan="3">Electric transfer</td><td colspan="4">Heat transfer</td></tr>
<tr><td></td><td colspan="3">Use of carbon nanotubes in electric cables and power lines for reduced power loss</td><td colspan="4">Carbon nanotube-based composites for optimized heat transfer</td></tr>
<tr><td>Energy storage</td><td colspan="4">Electric energy storage</td><td colspan="3">Chemical energy storage</td></tr>
<tr><td></td><td>Battery</td><td>supercapacitors</td><td colspan="2">Fuel cells</td><td colspan="2">hydrogen</td><td>Refining</td></tr>
<tr><td></td><td>Nanostructured electrodes for enhanced efficiency, and improved life in lithium-ion battery</td><td>Carbon nanotube, graphene, and nano-metal oxide-based electrodes for enhanced productivity</td><td colspan="2">Use of nano-sized platinum as a catalyst for improved performance</td><td colspan="2">Nano-porous materials for application in micro fuel cells</td><td>nano-catalysts for optimized fuel production (oil refining, desulphurization, coal liquefaction)</td></tr>
</table>

Similarly, other solar cells like dye solar cells and polymer solar cells have profited by using nanomaterials in terms of both efficiency and life span of the products. In the case of Gratzel dye-sensitized solar cells (DSSC), the use of nanocrystalline TiO_2, different nanostructured morphologies of TiO_2 like the nanotubes, nanorods, etc, have resulted in improved conversion power efficiencies due to their higher surface area, but also because of other properties like better photon absorption, reduced charge recombination as well as greater electron transport [4, 5]. Use of nanocomposites made of graphene and nano-TiO_2 as the photoanode helped in faster electron transport and reduced recombination resulting in enhanced power conversion efficiency of 40% [6, 7]. The use of quantum dots (graphene, bi-metallic nanoparticles, etc) has demonstrated advantages like enabling light-absorbing capacity in the infrared region, better photo chemical stability as compared to the use of dyes as sensitizers, and enhanced efficiency (approximately 42%) due to multiple electron generation [8].

Similarly, conversion of primary sources of energy to heat, electricity etc., requires high efficiency. Efficiency in steam power plants requires their operation at higher temperature and pressure; which requires heat and requires heat-resistant turbine materials. The use of carbon nanotube, ceramic, or inter-metallic coated turbine blades has been demonstrated in ensuring high resistance to heat and wear or tear; which has resulted in more efficient power plants. Coal-fired power plants, which convert chemical energy to electricity, are responsible for the large-scale generation of electricity but, at the same time, are major sources of greenhouse gas emissions. The use of nano-membranes for the capture of CO_2 during power generation in coal-fired power plants has enabled the environmental friendliness of the entire process. One of the major disadvantages in the commercialization of fuel cells is its high fabrication cost, inexpensive platinum metal catalyst, and limited supply of hydrogen used as a fuel. Also, the energy efficiency of fuel cells is 40–60% [3]. The use of nano-structured electrodes or membranes has boosted energy yield during the conversion of chemical energy in fuel cells, which has resulted in the economic application of such devices in automobiles, buildings, and the operation of mobile electronics [1]. The high electrical conductivity of carbon nanotubes has resulted in reduced power losses when applied to electric cables and power lines. Carbon nanotube-based composites have optimized heat transfer for use in industries and buildings.

In electric energy storage, nanotechnology has demonstrated immense potential, especially in lithium ion batteries and super-capacitors. It is well known that lithium ion battery is the major source of power and are used in various applications. Some of the major limitations, especially electrode materials in lithium ion batteries, are related to safety, environmental protection, cycle stability, high cost, and low power density. As per the latest reports, by using nano-structured electrode materials, the Li-ion rechargeable batteries have shown 10 folds improvement over their conventional counterpart in the following aspects: reduced size has brought about better conductivity; increased power, and hence the requirement of less time to recharge [9]. Nanomaterials have also improved battery life and have incorporated enhanced heat resistance in the electrodes. Work is in progress to commercialize nano-sized Li-ion batteries in hybrid and electric vehicles. Fuel cells have demonstrated good conversion efficiency of chemical energy to electricity. Due to its high efficiency, fast load response, and fuel use diversity, fuel cells are also used as energy storing devices. But the major demerits that are limiting the development and large-scale usage of fuel cells are high installation costs, use of expensive platinum catalysts, low battery power, etc. The use of nano-sized platinum as a catalyst has resulted in improvements in performance via more production of current (12 folds) and betterment in a lifetime (10-folds) as compared to conventional fuel cells [10, 11]. The use of nitrogen-doped carbon nanotubes in place of inexpensive platinum catalysts has resulted in an enhanced current generation (approximately 4 times) as compared to platinum-based fuel cells [12]. Supercapacitors have demonstrated high capacitance, power density, and better service life compared to lithium ion batteries, fuel cells, and traditional capacitors. This is mainly on account of revolutionary in the design and fabrication of electrode and electrolyte materials. Both carbon nanotubes and graphene have demonstrated better electrode materials for enhancing the productivity of supercapacitors. The reason cited is high surface area, higher mesoporosity, high electrical conductivity, and high charge transport capability [13, 14].

Thus, based on the superior characteristics of nanomaterials (high electron mobility, high permeability along with high electrical conductivity), there has been an increase in the demand for such materials for enhancing the performance of energy and storage devices, as evident from the analysis presented by the Grand View Research project over the forecast period of 2020 to 2027 [15]. The CAGR (compound annual growth rate) for the demand for nanomaterials (especially graphene) in the energy and power sector was predicted at 13.4%.

2.3 Nanomaterials and Sustainability Issues

In all energy and storage devices, nanotechnology has enabled the use of a wide range of nanomaterials ranging from carbon nanomaterials including carbon nanotubes, graphene, nanocellulose, metal oxide nanomaterials, nanocomposite coatings, and conducting polymer-based nano-composites. The important aspect is the precise control of various attributes of nanoparticles like particle size, pore size distribution, surface morphology, monodispersity, stability, etc during the synthesis of nanomaterials. This is because such attributes are directly related to the performance of energy devices.

Precise control of reaction conditions, capping agent, reducing agent, and solvent has led to the synthesis of monodisperse nanoparticles. Current methods employed for nanomaterials synthesis are via chemical vapor deposition, hydrothermal, sol-gel, co-precipitation, pyrolysis, etc These ensure better reaction rates and customized nanomaterials depending on their end-use [16].

Despite the advantages guaranteed in relation to particle attributes, such methods suffer from drawbacks related to their sustainability. Irrespective of the methods employed, nanomaterials prepared for use in semiconductors, electrodes, or any energy device require strict purity conditions as compared to bulk materials prepared by conventional methods. Despite the adherence to quality norms, the final product yield is low. In other words, the enormity in the quantity of waste generated is a burden on the environment. The third disadvantage is the requirement of repeated processing steps to achieve nanomaterials of desired quality intended for use in energy devices. For e.g. during the fabrication of wafers, cleaning requires their immersion in various chemical solvents followed by cleaning in ultrapure water. Since the last decade, cleaning has risen to approximately 100 executions per wafer [17]. Cleaning also warrants the use of toxic chemicals, surfactants, or ultrapure water. Some of the toxic chemicals used during the synthesis of 1D nanomaterials are volatile organic compounds (VOCs) e.g. methyl isobutyl ketone, tetrahydrofuran, N-methyl-2-pyrrolidone, propylene glycol monomethyl ether acetate, ethyl lactate, toluene, xylene, etc. and hazardous air pollutants (HAP) e.g. strong acids, arsine, phosphine, xylene, pyrene, etc [18–20]. There is also the requirement of ultrapure water for purification and cleaning. Nanomaterial synthesis also requires specialized environments like high temperature, vacuum conditions, ultrapure precursors, reagents, and solvents; all of which require high energy and additional purification costs [16, 21]. Such unsustainable synthesis methods result in high energy consumption, an excessive burden on the environment and economy. Literature studies reveal that the energy required for conversion of a kilogram of metallurgical grade silica to nano-silica for electronic use ranges from 110 to 210 KWatt hours [22].

Besides the toxic chemicals generated during the fabrication and processing of nanomaterials, the fate, transport, and lifetime of such nanomaterials and the associated risks have been certified by various life cycle assessment studies [23–25]. The other toxicity aspects of nanomaterials are related to their inefficient removal techniques; due to which there is a potential risk associated with their accidental leakage from wastes. Because of their high reactivity, they can be good adsorbers or carriers for other hazardous compounds. But the need for quantifying the environmental burden of the manufacturing or synthesis of nanomaterials has been explained in various works [26, 27].

Also, the adverse environmental effect of the synthesis of various nanomaterials has been demonstrated using the *E*-factor studies [28]. *E*-factor is a measure of environmental impact and sustainability by considering all materials and chemicals involved in the production process. Energy, water input, and the product of combustion are not included in the *E*-factor calculation. It is a material input (kg) ratio to product (kg). The study revealed that the current synthesis methods of nanomaterials under laboratory conditions produce 100,000 kg of waste per kg of product. As per the *E*-factor calculated, it was found that per kg of nanomaterials produced approximately 1000 times higher waste as compared to the per kg of bulk materials produced. Also, irrespective of the synthesis methods employed, the *E*-factor for nanomaterials synthesized with repeated purification steps was much higher than the same for nanomaterials developed without any purification.

Research on these negative implications of the fabrication and applications of nanomaterials has led to the development of new methodologies for the fabrication of green or sustainable nanomaterials that meet performance specifications and pose minimal environmental impact.

2.4 Green Nanomaterials Enhancing the Sustainability in Energy Applications

Adherence to the "12 Principles of Green Chemistry" during the fabrication has resulted in the development of green nanomaterials with subsequent sustainability in energy application. In the context of fabrication of nanomaterials, the principles of Green Chemistry as proposed by P. Anastas in the year 1998 [29] involves the use of water as a solvent wherever suitable, use of less toxic precursors, using lesser reagents, and as few as possible processing steps, using a reaction temperature as close as possible to room temperature and finally reducing the generation of toxic wastes and by-products. Thus, green nanotechnology deals with the design of environmentally benign nanoparticles using green reagents and processes and deals with minimizing the challenges due to the toxicity of nanoparticles to the environment.

The three principle reagents used during the fabrication of nanoparticles and have a controlling effect on the size, morphology, and monodispersity of the nanoparticles are the capping agent, reducing agent, and reaction solvent. Capping agents are added to the reaction medium to enhance nanoparticles' stability and hence property by preventing them from aggregating. The use of hetero-atom functionalized long-chain hydrocarbons (tri octyl-phosphine oxide, tri-n-octyl-phosphine, oleic acid, octadecylamine, etc oleyl amine, dodecyl amine, etc.) as capping agents have resulted in the fabrication of mono-dispersed nano-sized particles [30, 31]. The precise control of the size and morphology has enhanced the size-related properties and benefited the assembly of ordered patterns of nanoparticles for device fabrication. Chemical reduction is an important reaction during nanoparticle synthesis and the common reducing agents used to date are classified as hazardous. Examples of commonly used reducing agents are ethylene glycol, hydrazine, sodium borohydride, and formaldehyde [32]. The role of solvents during nanoparticle synthesis is to dissolve precursors, enable reaction medium, and disperse the as-formed nanoparticles. The majority of the solvents used are organic in nature (examples are ethanol, toluene, dimethyl formamide, etc.). The synthesis methods in practice involving the use of toxic hetero atom functionalized long-chain hydrocarbons, hazardous reducing agents, and toxic solvents are detrimental to the environment. Also, because of the strong bonding between the heteroatoms of the capping agent and nanoparticles, subsequent removal steps are tedious and time-consuming, thereby making the process more energy-intensive.

2.4.1 Green Reagents Used During Nanoparticle Synthesis

Because of the increased demand for sustainable energy devices, more emphasis has been placed on using greener reagents and processes during the fabrication of lab-scale nanoparticles and their subsequent scale-up [32]. About capping agents, the use of synthetic macromolecules like polymers (examples are polyvinyl pyrrolidone, polyvinyl alcohol, polyethylene glycol, polypropylene glycol, polyacrylic acid, and polyphenylene oxide) and dendrimers (examples are poly(amidoamine) and poly(propylene-imine)) have resulted in the synthesis of specific nano-sized particles; but because of the weak interactions involved, capped nanoparticles are easily removed; thereby the requirement of long removal steps are negligible [33, 34]. More recently, the use of polysaccharides, bio-molecules, and other small molecules has increased during the synthesis of nanoparticles. Polysaccharides are polymers having biological origins and hence are biocompatible. Also, because of their mild interactions with the nanoparticles, their usage as capping agents has resulted in the synthesis of environmentally benign nanoparticles requiring only water as the solvent. Polysaccharides are also classic examples of green reducing agents. The presence of infinite hydroxyl groups in the repeated glucose units promotes association, thereby enabling aggregation, complexation, and enhanced reduction of metal/metal oxide/semiconductor precursors, which promotes the formation of nanoparticles [35]. Because of the good capping properties of such polysaccharides, synthesized nanoparticles are well dispersed in an aqueous medium without particle aggregation. Examples of polysaccharides used to date to synthesize green nanoparticles are starch, cellulose, chitin, and dextran. In a study conducted by Liu et al. [36], monodisperse Au nanoparticles of an average diameter of 8.2nm were synthesized using green reagents like β-d-glucose which acted as a capping and reducing agent. The method required less reaction time and fewer reagents for bringing about nucleation and growth of nanoparticles. The method required adjustment in aqueous pH wherein the reaction kinetics for reducing Au^{3+} was precisely controlled. In the study [35], the use of green reagents like starch (as a capping agent) and β-d-glucose (as a reducing agent) resulted in the synthesis of Ag nanoparticles in an aqueous medium under a reaction temperature of 40 °C and a long reaction time of 20hrs. But the nanoparticles with an average diameter of 5.3nm demonstrated high stability and monodispersity in aqueous suspension.

Cellulose extracted from various tree/plant species has also been used as a capping agent and reducing agent for nanoparticle synthesis. In a study conducted by Mochochoko et al. [37], cellulose extracted from an environmentally problematic aquatic weed (water hyacinth) was used as a capping and reducing agent for the green synthesis of Ag nanoparticles. The average particle diameter ranged from 2-5nm depending on the aqueous pH. Ag nanoparticles were synthesized by a green method using seed extract of *Hibiscus sabdariffa* as the reducing agent. Cellulose extracted from the stem of the same plant species was used to support the nanoparticles [38]. The cellulose-supported Ag nanoparticles were evenly and well dispersed, demonstrating an average size of 4 nm.

Bio-molecules like peptides and proteins have promoted the stability as well as the morphological properties of different nanoparticles during synthesis [39–41]. In a study [42], Ag nanoparticles were synthesized using green reagents (cell-free extract of phytopathogenic soil-borne fungus *Macrophomina phaseolina* (Tassi) Goid). The protein-capped Ag nanoparticles demonstrated spherical morphology size having a size range of 5 to 40 nm (majority exhibiting 16 to 20 nm in diameter). Protein extracted extracellularly from *Escherichia coli* served both as a capping as well as a reducing agent during the eco-friendly synthesis of gold and silver nanoparticles under ambient conditions [43]. Literature reveals the success of using different bio-molecules extracted extracellularly from different microorganisms (like algae, bacteria, and fungi) during the green synthesis of Ag nanoparticles [44–47]. The involvement of various other phytochemicals (polyphenols, alkaloids, flavonoids, sugars, terpenoids, etc.) extracted from various plant species has been shown to directly involve the synthesis of nanoparticles [48–56]. Their involvement has been in the form of bio-reduction, stabilization, and capping agents. Besides being biocompatible, environmentally friendly capping agents for nanomaterial synthesis, the bio-molecules provide sufficient adsorption sites to induce higher reaction rates and enhance the growth of nanoparticles in different directions. Thus, the use of bio-molecules during the synthesis of nanoparticles has demonstrated environmentally friendly options and has executed a greater and precise control over the morphology of nanoparticles thus synthesized.

The use of small molecules like formaldehyde, oxalate, CO, etc., as capping agents has enabled nanoparticle synthesis with no subsequent requirement of removal steps thereby making the process less prone to environmental risks and less energy-intensive as compared to the use of conventional capping agents [57, 58].

The environmentally friendly options available for solvent selection are water, supercritical fluids, and ionic liquids [32]. The use of water has been cited to be environmentally favorable as per the Green Chemistry principles [29]. This is because of its non-toxicity and non-flammable nature. Also, because of its easy and wider availability, water use as a solvent is economically feasible. Supercritical fluids are also non-toxic, non-inflammable, and inexpensive but are energetically superior compared to the use of water as a solvent [59, 60]. At critical temperatures and pressure of 646 K and 221 bar, supercritical water has exhibited enhanced solubility towards non-polar organic solvents due to its reduced dielectric constant. Also, in a study, it was demonstrated that around the supercritical point, the solubility of metal oxides in water showed a decrease to approach the degree of supersaturation faster, resulting in the formation of a large number of small-sized nanoparticles [61]. Another study demonstrated that the use of supercritical water as the solvent facilitated the formation of nanoparticles with narrow size dimensions and the addition of different organic solvents to the reaction medium caused enhanced dispersion depending on the particular application [62, 63]. At critical conditions, the reduced dielectric constant of water caused increased solubility of the organic solvents. As a result of which, complexation between the organics and surface of nanomaterials was facilitated, resulting in increased dispersibility.

Supercritical CO_2 is another example of a green solvent used in nanoparticle synthesis [59]. But its use is more energetically favorable than water because of the lower temperature and pressure requirements (304 K and 74 bar respectively) [63]. Also, supercritical CO_2 can be recovered and reused via depressurization; making the process more favorable for the industrial production of nanoparticles. By manipulating the temperature and pressure, the physical properties of CO_2 like phase, density, viscosity, and dielectric constant could be altered and thereby result in controlled synthesis of nanoparticles with respect to size distribution, morphology, and porosity.

Ionic liquids are another class of green solvents that demonstrate lower toxicity and lower energy requirements during nanoparticle synthesis than conventional organic solvents [63]. Also, the requirement of capping agents are nullified, thereby reducing the material requirements.

2.4.2 Green Processes Involved in Nanoparticle Synthesis

The synthesis of green nanomaterials requires green reagents and chemicals and involves green methodologies and process conditions. The conventional synthetic processes involve energy input via external heating sources (heating mantle, water or oil bath, or furnace). But such heating procedures require higher energy input as well as involve high reaction temperature and longer reaction times. Hence such processes are not efficient with respect to efficiency and energy considerations. Non-conventional energy input via the use of microwave and ultrasound has resulted in nanoparticle synthesis with greater control on its size and morphology and has immense environmental benefits. Microwave heating carried out in the presence of dimethylsulfoxide (DMSO) and dimethylformamide (DMF), has brought about higher heating rates and reduced reaction time [64]. Both DMSO and DMF are highly polar solvents and absorb microwave irradiation to bring about localized superheating effects. To extend the benefit of microwave irradiation to a range of non-polar solvents, the addition of ionic liquids to the reaction medium has promoted the heating effects without affecting the nanoparticle formation. Ionic liquids are highly polar solvents and are good microwave absorbers that bring about fast and homogenous heating in the reaction system. Besides the environmental benefits, controlling microwave heating and degree of penetration via its applied wavelength strongly influences nanoparticle nucleation and growth. Small-sized nanoparticles with narrow size distribution were attained under the intense heating rate provided by microwave irradiation. Similarly, larger-sized nanoparticles were the result of prolonged microwave irradiation. A fast and green method employing microwave heating assistance was used to synthesize high entropy (Mg, Cu, Ni, Co, Zn) oxide nanoparticles [65]. The method not only demonstrated faster kinetics, employing low temperature but resulted in the fabrication of small-sized nanoparticles (size distribution of 20–70 nm with an average particle size of 44 nm). The metal oxide nanoparticles demonstrated significant storage properties and excellent stability when used as an anodic material in Li-ion batteries. In yet another study, the microwave-assisted co-precipitation method was used for the synthesis of nanoparticles of ZnS using cetyltrimethylammonium bromide as the capping agent [66]. The as-prepared nanoparticles were of low dimension (crystallite size of ~3 nm). Also, the high energy gap of 3.59–3.64 eV, high dielectric constant of 27 and 35, and low loss values reveal the suitability of the nanoparticles for use in optoelectronics. Green carbon-supported ultra-small nanoparticles of ternary alloy (Pt-Sn-Rh) were successfully fabricated using the microwave heating method [67]. The particles synthesized had a uniform size distribution and a particle size of 2nm. The synthesis method was fast bringing about uniform heating, thereby lowering the cost of production, and was energy efficient. The nanomaterial demonstrated as an electrocatalyst having long-term durability; exhibiting enhanced mass activities for the ethanol oxidation reaction (EOR) and methanol oxidation reaction (MOR). A comparative study showed the performance of the electrocatalyst for EOR and MOR was 5.74 and 3.70 times higher than the commercial Pt/C catalysts.

Nanoparticle synthesis using ultrasonication has demonstrated faster kinetics and has required ambient reaction conditions without the need for reducing agents [68, 69]. Also, the solvent used during ultrasonication is water which is considered non-toxic and green [70]. The ultrasound technique is hence considered green for the synthesis of nanoparticles. In comparison to microwaves, the heating is not generated directly from the ultrasound, but acoustic cavitation is generated under the influence of the acoustic energy. As a result of the cavitation, high temperatures (2000-10,000 K) and high pressures (100-1000 MPa) are generated, promoting nanoparticle synthesis. In a study conducted by Murcia et al., [71] ultrasonication technique was used to synthesize Cd-Se-ZnS core-shell quantum dots (QD). The technique resulted in a high yield of QD to the tune of 50-60% and synthesis of narrow-sized nanoparticles (approximately 10%), which demonstrated high luminescence properties. Besides resulting in good yields, ultrasonication could bring about the synthesis of the QDs at temperatures much lower than that required during conventional heating procedures. Ultrasonic assisted green synthesis method resulted in the fabrication of small sized Pt-nanoparticles using different solvents (water, ethanol, and methanol) and extracts from *Prosopis farcta* fruits as reducing agents [72]. The nanoparticles exhibited semi-spherical morphology having a size of 1.6-5 nm, but the maximum stability was attained with ethanol as the solvent.

2.4.3 Biomass Based Green Nanotechnology in Energy Devices

Porous carbon nanomaterials derived from biomass have demonstrated suitable electrode materials in various energy storage devices like lithium-ion/sodium-ion batteries and supercapacitors [73]. It is a well-known fact that carbon materials like activated carbons, graphene, and carbon nanotubes exhibit high specific surface area, good pore size distribution, adjustable pore size, stable, chemical properties, and can be tuned to incorporate various hetero atom-based functional groups. Because of these features, carbon-based porous materials, when used as electrodes, have demonstrated better electrical conductivity, cycle stability, and electric charge storage capability in both lithium-ion batteries and supercapacitors as compared to the use of conductive polymers or metal oxides as electrodes. While activated carbon, graphene, and carbon nanotubes are not cost-effective, the other demerit of activated carbon is its relatively low specific surface area. In this context, biomass-derived porous carbon nanomaterials synthesized via green processes like one-step chemical activation, microwave, hydrothermal carbonization, microwave-assisted chemical activation, etc., have served the dual purpose of being economical and clean/green electrode materials. The other advantages related to its Physico-chemical characteristics are the inherent presence of hetero-atoms and natural functional groups, which tend to enhance the capacitance of carbon electrodes. Widely distributed porous networks help to improve the transmission of charged ions leading to significantly improved electrical conductivity. Coal tar pitch was used as a precursor, which when subjected to microwave (600W, 30 mins) assisted chemical activation using KOH as the activator and MgO as the template resulting in a carbon material having high S_{BET} (1394m^2/g) and pore volume as compared to those prepared by conventional heating [74]. When used as an electrode in 6M aqueous electrolyte, the specific capacitance of the supercapacitor was recorded at 224F/g. The synthesis protocol adopted was green in the sense that a single step using microwave technology, less time, and fewer reagents were involved. Single-step pyrolysis (1000 °C for 30mins) of dead Neem plant leaves was used to fabricate a porous carbon having an S_{BET} of 1230m^2/g. The charge storage capacity of the cell developed using the fabricated carbon as the electrode and using an aqueous inorganic electrolyte (1M H_2SO_4) was 400 F/g and an energy density of 55 Wh/kg [75]. The graphene-CNT-based electrode in the presence of 0.5M H_2SO_4 exhibited a maximum energy density of 21.74 Wh/kg [76]. A similar fabrication step (carbonization at 600 °C for 3 hrs) resulted in a porous carbon (S_{BET} of 746 m^2/g) from seaweeds. The carbon used as an electrode demonstrated a maximum capacitance of 264 F/g using 1M H_2SO_4 as an aqueous inorganic electrolyte [77]. Cotton cellulose was impregnated with a pore creator ($Mg(NO_3)_2$) and subjected to pyrolysis at 800 °C for 2hrs to develop a porous carbon having an S_{BET} of 1260m^2/g [78]. When used as an anode in a lithium-ion battery, it exhibited promising electrochemical behavior showing a capacity of 793mAh/g at a current density of 0.5 A/g after 500 cycles of operation.

Results of the performance of various green carbon-based nano porous materials on the efficiency of supercapacitors/lithium/sodium-ion batteries are summarized in Table-2.2.

Table 2.2 Performance of various biomass-derived carbon-based electrodes in energy storing device

Biomass-derived carbon electrode	Synthesis steps	Energy device	Capacitance	Other features	Reference
Coal tar pitch	Microwave-assisted chemical activation with KOH + MgO template	supercapacitor	241F/g in 6M KOH	SSA: 1143m^2/g	[74]
Neem dried leaves	Pyrolysis at 1000 °C for 30mins	supercapacitors	400F/g in 1M H_2SO_4	SSA: 1230m^2/g	[75]
Seaweed	Pyrolysis at 1000 °C for 30mins	supercapacitors	264F/g in 1M H_2SO_4	SSA: 746m^2/g	[76]

(*Continued*)

Table 2.2 Continued

Biomass-derived carbon electrode	Synthesis steps	Energy device	Capacitance	Other features	Reference
Cotton cellulose	Template method using $Mg(NO_3)_2$ as template	Lithium-ion battery	793mAh/g	SSA: 1260m^2/g	[78]
Sunflower seed shell	Chemical activation with KOH	supercapacitors	311F/g in 30wt % KOH	SSA: 2509m^2/g	[79]
Banana fibers	Chemical activation with $ZnCl_2$	supercapacitors	74F/g in 1M Na_2SO_4	SSA: 1097m^2/g	[80]
Rice husk	Chemical activation with NaOH (750 °C for 30mins)	supercapacitors	210F/g in 3M KCl	SSA: 1886m^2/g	[81]
Table sugar	Pyrolysis at 1050 °C	Lithium-ion battery	650mAh/g	NA	[82]
Coconut oil derived CNP	incineration	Sodium-ion battery	741mAh/g	NA	[82]
Rubber tube derived carbon soot	Controlled oxidation	Lithium-ion battery	190mAh/g	NA	[83]
Walnut shell derived carbon nanofibers	electrospinning	Lithium-ion battery	380mAh/g	NA	[84]
Shaddock/pomelo peels	Chemical activation with H_3PO_4 (700 °C)	Sodium-ion battery	314.5mAh/g	SSA: 1272m^2/g	[85]

2.5 Conclusion

The last decade has seen a rising awareness of greener synthetic approaches for nanomaterials to develop sustainable energy devices. Research on the use of nanomaterials at the laboratory level and scale up-operations has seen a boost in the performance of various energy devices like solar cells, fuel cells, batteries, supercapacitors, etc. As a result of which, the demand for nanomaterials has seen an upsurge. But, the conventional synthesis method employed using toxic reagents, lengthy procedures, and generating toxic wastes have raised sustainability issues related to nanoparticle synthesis and nanomaterials fabrication. The adoption of green nanoparticle synthesis protocols has paved the pathway for safer nanomaterials with the desired morphological and Physico-chemical features. However, there is still the challenge of the potentially toxic effects of nanoparticles on the environment. The application of green nanomaterials has been reported in various applications in different areas of energy. A schematic representation (Figure 2.1) has thus summarized the different green processes/reagents/reaction conditions involved in the synthesis of green nanomaterials from biobased precursors and the sustainable applications thereof in different areas of energy. But the sustainable applications of green nanomaterials are still in their infant stage. Challenges still exist related to the fate of such nanoparticles during their

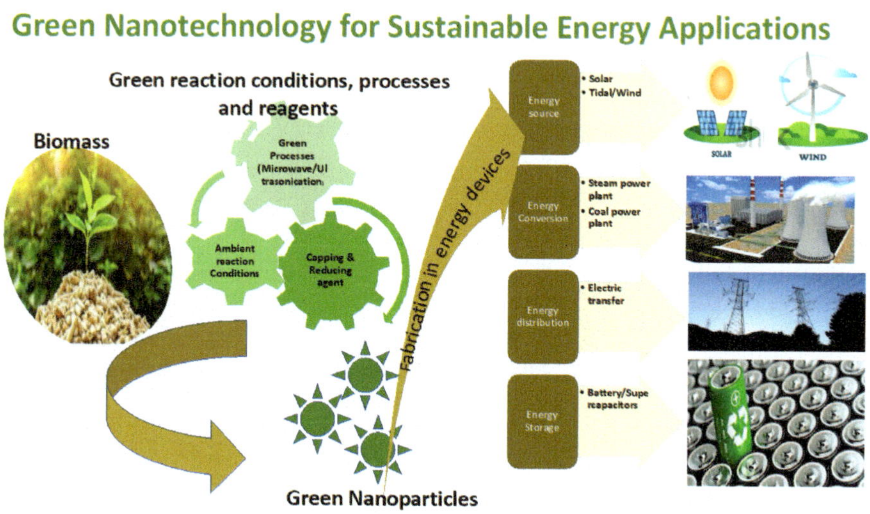

Figure 2.1 Schematic representation of green nanotechnology for Sustainable energy Application

application and usage. There are issues related to the undesirable effects on people having long-term exposure to nanomaterials. More research and analysis on such risk factors need to be undertaken to have information on safety measures that need to be taken for overall hazard management. The application of green nanotechnology in the energy sector thus involves the fabrication of green nanomaterials and requires adequate study and analysis of the effect of such technology on occupation health security. Only then can the technology have wider acceptance and adoption and be marketed to generate a green economy.

Acknowledgments

This work is supported by the Graphic Era Deemed to be University, Dehradun, Uttarakhand.

References

[1] Luther W. Application of Nanotechnologies in the Energy Sector. Germany: Hessian Ministry of Economy, Transport, Urban and Regional Development, 2008. HA Hessen Agentur GmbH Abraham-Lincoln-StraSSe 38-42 65189 Wiesbaden, Germany. https://www.technologieland-hessen.de/mm/NanoEnergy_web.pdf

[2] Joachim C. To be nano or not to be nano? Nat. Mater. 2005; 4 (2): 107–109. DOI: https://doi.org/10.1038/nmat1319

[3] Christian F, Edith, Selly, Adityawarman D, Indarto A. Application of nanotechnologies in the energy sector: A brief and short review. Front. Energy. 2013; 7: 6–18. DOI: https://doi.org/10.1007/s11708-012-0219-5

[4] Mor GK, Shankar K, Paulose M, Varghese OK, Grimes CA. Use of highly-ordered TiO(2) nanotube arrays in dye-sensitized solar cells. Nano Lett. 2006; 6: 215-218. DOI: https://doi.org/10.1021/nl052099j

[5] Paulose M, Shankar K, Varghese OK, Mor GK, Grimes CA. Application of highly-ordered TiO_2 nanotube-arrays in heterojunction dye-sensitized solar cells, J. Phys. D: Appl. Phys. 2006; 39: 2498. https://iopscience.iop.org/article/10.1088/0022-3727/39/12/005

[6] Yang N, Zhai J, Wang D, Chen Y, Jiang L. Two-dimensional graphene bridges enhanced photoinduced charge transport in dye-sensitized solar cells. ACS Nano, 2010; 4: 887–894. DOI: https://doi.org/10.1021/nn901660v

[7] Yang Y, Qu L, Dai L, Kang TS, Durstock M. Electrophoresis Coating of Titanium Dioxide on Aligned Carbon Nanotubes for Controlled Syntheses of Photoelectronic Nanomaterials. Adv. Mater. (2007;19: 1239-1243. DOI: https://doi.org/10.1002/adma.200602181

[8] Kamat PV. Quantum dot solar cells. The next big thing in photovoltaics. J. Phys. Chem. Lett. 2013; 4: 908–918. DOI: https://doi.org/10.1021/jz400052e

[9] Liang S, Zhu X, Lian P, Yang W, Wang H. Superior cycle performance of Sn@C/graphene nanocomposite as an anode material for lithium-ion batteries. J. Solid State Chem. 2011;184(6): 1400–1404. URL: http://www.scut.hhwang.ycym.com/Uploads/201703/58c926f854c76.pdf

[10] Antolini E, Perez J. The renaissance of unsupported nanostructured catalysts for low-temperature fuel cells: from the size to the shape of metal nanostructures. J. Mater Sci, 2011; 46(13): 4435–4457. DOI: https://doi.org/10.1007/s10853-011-5499-3

[11] Guo S, Sun S. FePt nanoparticles assembled on graphene as an enhanced catalyst for the oxygen reduction reaction. J.Am . Chem Soc. 2012; 134(5): 2492–2495. DOI: https://doi.org/10.1021/ja2104334

[12] Dai L, Chang DW, Baek JB, Lu W. Carbon nanomaterials for advanced energy conversion and storage. Small, 2012; 8(8): 1130–1166. DOI: https://doi.org/10.1002/smll.201101594

[13] Niu C, Sichel EK, Hoch R, Moy D, Tennent H. High power electrochemical capacitors based on carbon nanotube electrodes. Appl. Phys. Lett.1997; 70: 1480-1482. DOI: https://doi.org/10.1063/1.118568

[14] Yang W, Ni M, Ren X. Graphene in supercapacitor applications. Curr. Opin. Colloid Interface Sci. 2015; 20: 416–428. Yang W, Ni M, Ren X. Graphene in supercapacitor applications. Curr. Opin. Colloid Interface Sci. 2015; 20: 416–428. DOI: https://doi.org/10.1016/j.cocis.2015.10.009

[15] Nanomaterials Market Size, Share & Trends Analysis Report By Product (Carbon Nanotubes, Titanium Dioxide), By Application (Medical, Electronics, Paints & Coatings), By Region, And Segment Forecasts, 2020 – 2027. https://www.grandviewresearch.com/industry-analysis/nanotechnology-and-nanomaterials-market

[16] Şengül H, Theis TL, Ghosh S. Toward Sustainable Nanoproducts. J. Ind. Ecol. 2008; 12(3), 329–359. DOI: https://doi.org/10.1111/j.1530-9290.2008.00046.x

[17] Verhaverbeke S, Kuppurao S, Beaudry C, Truman JK. Single-wafer, short cycle time wet clean technology. Semiconductor International 2002; 25(8): 91. https://www.researchgate.net/publication/279575669_Single-wafer_short_cycle_time_wet_clean_technology

[18] Lai HJ, Lin MCC, Yang MH, Li AK. Synthesis of carbon nanotubes using polycyclic aromatic hydrocarbons as carbon sources in an arc discharge. Materials Science & Engineering C: Biomimetic and Supramolecular Systems 2001;16(1–2): 23–26. DOI: https://doi.org/10.1016/S0928-4931(01)00303-4

[19] Hoppe H, Niggemann MC, Winder J, Kraut R, Hiesgen A, Hinsch D, Meissner, Sariciftci NS. Nanoscale morphology of conjugated polymer/fullerene-based bulk-heterojunction solar cells. Adv. Funct. Mater. 2004; 14(10): 1005–1011. DOI: https://doi.org/10.1002/adfm.200305026

[20] Mukherjee B, Ravishankar N. A novel solvothermal method for nanoparticle thin films and coatings. Nanotechnology 2007; 18(2): Art. No. 025603. DOI: http://dx.doi.org/10.1088/0957-4484/18/2/025603

[21] Javey A, Dai HJ. Regular arrays of 2 nm metal nanoparticles for deterministic synthesis of nanomaterials. J. Am. Chem. Soc. 2005; 127(34): 11942–11943. DOI: https://doi.org/10.1021/ja0536668

[22] Bartolozzi I, Daddi T, Punta C, Fiorati A, Iraldo F. Life cycle assessment of emerging environmental technologies in the early stage of development: A case study on nanostructured materials, J. Ind. Ecol. 2019; 24: 1, 101-115. DOI: https://doi.org/10.1111/jiec.12959

[23] Phylipsen, G. J. M., Alsema, EA. (1995). Environmental life-cycle assessment of multi crystalline silicon solar cell modules. Utrecht University report no. 95057. www.chem.uu.nl/nws/ www/publica/95057.pdf. Accessed January 2007.

[24] Gallagher MJ, Allen C, Buchman JT, Qiu TA, Clement PL, Krause MOP, Gilbertson LM. Research highlights: applications of life-cycle assessment as a tool for characterizing environmental impacts of engineered nanomaterials. Environ. Sci. Nano. 2017; 4: 276–281. DOI: https://doi.org/10.1039/C7EN90005H

[25] Tsang MP, Kikuchi-Uehara E, Sonnemann GW, Aymonier C, Hirao M. Evaluating nanotechnology opportunities and risks through integration of life-cycle and risk assessment. Nat. Nanotechnol. 2017; 12: 734–739. DOI: https://doi.org/10.1038/nnano.2017.132

[26] Romero-Franco M, Godwin HA, Bilal M, Cohen Y. Needs and challenges for assessing the environmental impacts of engineered nanomaterials (enms). Beilstein J. Nanotechnol. 2017; 8: 989–1014. DOI: https://doi.org/10.3762/bjnano.8.101

[27] Hutchison JE. Greener nanoscience: a proactive approach to advancing applications and reducing implications of nanotechnology. ACS Nano. 2008; 2: 395–402. DOI: https://doi.org/10.1021/nn800131j

[28] Eckelman MJ, Zimmerman JB, Anastas PT. Toward green nano: E-factor analysis of several nanomaterial syntheses. J. Ind. Ecol. 2008; 12: 316–328. DOI: https://doi.org/10.1111/j.1530-9290.2008.00043.x

[29] Anastas, P. T., Warner, J. C. (1998). Green chemistry: Theory and Practice, Oxford University Press Oxford. https://global.oup.com/academic/product/green-chemistry-9780198506980?cc=us&lang=en&

[30] Wang X, Zhuang J, Peng Q, Li Y. A general strategy for nanocrystal synthesis Nature. 2005; 437: 121–124. DOI: https://doi.org/10.1038/nature03968

[31] Park J, Joo J, Kwon SG, Jang Y, Hyeon T. Synthesis of monodisperse spherical nanocrystals, Angew. Chem. Int. Ed. 2007; 46: 4630–4660. DOI: https://doi.org/10.1002/anie.200603148

[32] Duan H, Wang D, Li Y. Green chemistry for nanoparticle synthesis. Chem. Soc. Review. 2015; 44(1): 5778–5792. DOI: https://doi.org/10.1039/c4cs00363b

[33] Astruc D, Lu F, Aranzaes JR. Nanoparticles as recyclable catalysts: the frontier between homogeneous and heterogeneous catalysis. Angew. Chem. Int. Ed. 2005; 44: 7852–7872. DOI: https://doi.org/10.1002/anie.200500766

[34] Lopez-Sanchez JA, Dimitratos N, Hammond C, Brett GL, Kesavan L, White S, Miedziak P, Tiruvalam R, Jenkins RL, Carley AF, Knight D, Kiely CJ, and Hutchings GJ, Facile removal of stabilizer-ligands from supported gold nanoparticles. Nat. Chem. 2011; 3: 551–556. DOI: https://doi.org/10.1038/nchem.1066

[35] Raveendran P, Fu J, Wallen SL. Completely "green" synthesis and stabilization of metal nanoparticles. J. Am. Chem. Soc. 2003; 125: 13940–13941. DOI: https://doi.org/10.1021/ja029267j

[36] Liu J, Qin G, Raveendran P, Ikushima Y. Facile "Green" Synthesis, Characterization, and Catalytic Function of β-D-Glucose-Stabilized Au Nanocrystals. Chemistry - A European Journal, 2006;12(8): 2131–2138. DOI: https://doi.org/10.1002/chem.200500925

[37] Mochochoko T, Oluwafemi OS, Jumbam DN, Songca SP. Green synthesis of silver nanoparticles using cellulose extracted from an aquatic weed; water hyacinth. Carbohydr Polym. (2013;98(1): 290–294. DOI: https://doi.org/10.1016/j.carbpol.2013.05.038

[38] Goswami M, Baruah D, Das AM. Green synthesis of silver nanoparticles supported cellulose and their catalytic application in the scavenging of organic dyes. New Journal of Chemistry, (2018; 42(13): 10868–10878. DOI: https://doi.org/10.1039/c8nj00526e

[39] Chin-Yi Chiu YL, Ruan L, Ye X, Murray CB, Huang Y. Platinum nanocrystals selectively shaped using facet-specific peptide sequences. Nat. Chem. 2011; 3: 393–399. DOI: https://doi.org/10.1038/nchem.1025

[40] Lévy R. Peptide-Capped Gold Nanoparticles: Towards Artificial Proteins. Chem.Bio. Chem. 2006; 7(8): 1141–1145. DOI: https://doi.org/10.1002/cbic.200600129

[41] Sanghi R, Verma P. Biomimetic synthesis and characterization of protein capped silver nanoparticles. Bioresour. Technol. 2009; 100(1): 501-504. DOI: https://doi.org/10.1016/j.biortech.2008.05.048

[42] Chowdhury S, Basu A, Kundu S. Green synthesis of protein capped silver nanoparticles from phytopathogenic fungus *Macrophomina phaseolina* (Tassi) *Goid* with antimicrobial properties against multidrug-resistant bacteria. Nanoscale Res. Lett. (2014; 9: 365. DOI: https://doi.org/10.1186/1556-276X-9-365

[43] Veeraapandian S, Sawant S N, Doble M. Antibacterial and Antioxidant Activity of Protein Capped Silver and Gold Nanoparticles Synthesized with Escherichia coli. J. Biomedi. Nanotechnol. 2012; 8(1): 140–148. DOI: https://doi.org/10.1166/jbn.2012.1356

[44] Sudha SS, Rajamanickam K, Rengaramanujam J. Microalgae mediated synthesis of silver nanoparticles and their antibacterial activity against pathogenic bacteria. Ind. J. Expt. Biol. 2013; 51: 393–399. http://nopr.niscair.res.in/bitstream/123456789/17672/1/IJEB%2051(5)%20393-399.pdf

[45] Ganeshkumar C, Mamidyala SK. Extracellular synthesis of silver nanoparticles using culture supernatant of *Pseudomonas aeruginosa* . Colloids Surf B: Biointerfaces 2011; 84: 462–466. DOI: https://doi.org/10.1016/j.colsurfb.2011.01.042

[46] Shivaji S, Madhu S, Singh S. Extracellular synthesis of antibacterial silver nanoparticles using psychrophilic bacteria. Process Biochem. 2011; 46: 1800–1807. DOI: https://doi.org/10.1016/j.procbio.2011.06.008

[47] Vahabi K, Mansoori GA, Karimi S. Biosynthesis of silver nanoparticles by fungus Trichoderma reesei (a route for large-scale production of AgNPs). Insci J. 2011; 1: 65–79. DOI: https://doi.org/10.5640/insc.010165

[48] Uma Suganya KS, Govindaraju K, Ganesh Kumar V, Dhas ST, Karthick V, Singaravelu G, Elanchezhiyan M. Blue green alga mediated synthesis of gold nanoparticles and its antibacterial efficacy against Gram positive organisms. Mater. Sci. Eng. C. 2015; 47: 351–356. DOI: https://doi.org/10.1016/j.msec.2014.11.043

[49] Jayaprakash N, Vijaya JJ, Kaviyarasu K, Kombaiah K, Kennedy LJ, Ramalingam RJ, Munusamy MA, Al-Lohedan HA. Green synthesis of Ag nanoparticles using tamarind fruit extract for the antibacterial studies. J. Photochem. Photobiol. B 2017; 169: 178-85. DOI: https://doi.org/10.1016/j.jphotobiol.2017.03.013

[50] Khodadadi B, Bordbar M, Nasrollahzadeh M. Green synthesis of pd nanoparticles at apricot kernel shell substrate using Salvia hydrangea extract: Catalytic activity for reduction of organic dyes. J. Colloid Interface Sci. 2017; 490: 1-10. DOI: https://doi.org/10.1016/j.jcis.2016.11.032

[51] Marslin G, Siram K, Maqbool Q, Selvakesavan RK, Kruszka D, Kachlicki P, Franklin G. Secondary metabolites in the green synthesis of metallic nanoparticles. Materials (Basel) 2018;11(6): 940. DOI: https://doi.org/10.3390/ma11060940

[52] Jain S, Mehata MS. Medicinal plant leaf extract and pure flavonoid mediated green synthesis of silver nanoparticles and their enhanced antibacterial property. Sci. Rep. 2017; 7: 15867. DOI: https://doi.org/10.1038/s41598-017-15724-8

[53] Kumar V, Yadav SK. Plant-mediated synthesis of silver and gold nanoparticles and their applications. J. Chem. Biotechnol. 2009; 84:151-7. DOI: https://doi.org/10.1002/jctb.2023

[54] Ahmed S, Ahmad M, Swami BL, Ikram S. A review on plants extract mediated synthesis of silver nanoparticles for antimicrobial applications: A green expertise. J. Adv. Res. 2016; 7: 17-28. DOI: https://doi.org/10.1016/j.jare.2015.02.007

[55] Zuas O, Hamim N, Sampora Y. Bio-synthesis of silver nanoparticles using water extract of *Myrmecodia pendan* (sarang semut plant). Mater. Lett. 2014; 123: 156-9. DOI: https://doi.org/10.1016/j.matlet.2014.03.026

[56] Muniyappan N, Nagarajan NS. Green synthesis of silver nanoparticles with Dalbergia spinosa leaves and their applications in biological and catalytic activities. Process Bio. Chem. 2014; 49: 1054-61. https://agris.fao.org/agris-search/search.do?recordID=US201700124490

[57] You H, Yang S, Ding B, Yang H. Synthesis of colloidal metal and metal alloy nanoparticles for electrochemical energy applications. Chem. Soc. Rev., (2013; 42(7): 2880–2904. DOI: https://doi.org/10.1039/c2cs35319a

[58] Zhang C, Hwang SY, Trout A, Peng Z. Solid-State Chemistry-Enabled Scalable Production of Octahedral Pt–Ni Alloy Electrocatalyst for Oxygen Reduction Reaction. J. Am. Chem. Soc., 2014, 136, 7805–7808. DOI: https://doi.org/10.1021/ja501293x

[59] Johnston KP, Shah PS. Making Nanoscale Materials with Supercritical Fluids. Mater. Sci. 2004; 303: 482–483. DOI: https://doi.org/10.1126/science.1093951

[60] Pollet P, Davey EA, Urenã-Benavides EE, Eckert CA, Liotta CL. Solvents for sustainable chemical processes. Green Chem., 2014; 16: 1034. DOI: https://doi.org/10.1039/C3GC42302F

[61] Sue K, Adschiri T, Arai K. Predictive Model for Equilibrium Constants of Aqueous Inorganic Species at Subcritical and Supercritical Conditions. Ind. Eng. Chem. Res., 2002; 41: 3298–3306. DOI: https://doi.org/10.1021/ie010956y

[62] Adschiri T, Lee YW, Goto M, Takami S. Green materials synthesis with supercritical water. Green Chem., 2011; 13: 1380. DOI: https://doi.org/10.1039/C1GC15158D

[63] DeSimone JM. Practical approaches to green solvents. Sci., 2002; 297: 799–803. DOI: https://doi.org/10.1126/science.1069622

[64] Baghbanzadeh M, Carbone L, Cozzoli PD. Kappe CO. Microwave-Assisted Synthesis of Colloidal Inorganic Nanocrystals. Angew. Chem., Int. Ed., 2011; 50: 11312–11359. DOI: https://doi.org/10.1002/anie.201101274

[65] Kheradmandfard M, Minouei H, Tsvetkov N, Vayghan AK, Kashani-Bozorg SF, Kim G, Hong SI, Kim DE. Ultrafast green microwave-assisted synthesis of high-entropy oxide nanoparticles for Li-ion battery applications, Mater. Chem. Phys. 2021; 262: 124265. DOI: https://doi.org/10.1016/j.matchemphys.2021.124265

[66] Jubeera EM, Manthrammel MA, Shkir M, Subha PA, Yahia IS, Alfaify SA. Microwave assisted synthesis of quantum dots like ZnS nanoparticles for optoelectronic applications: An effect

of CTAB concentrations, Optik. 2021; 240: 166812. DOI: https://doi.org/10.1016/j.ijleo.2021.166812

[67] Hu X, Song P, Yang X, Wang C, Wang J, Tang Y, Zhang J, Mao, Z. One-step microwave-assisted synthesis of carbon-supported ternary Pt-Sn-Rh alloy nanoparticles for fuel cells. Journal of the Taiwan Institute of Chemical Engineers. (2020); 115: 272-278. DOI: https://doi.org/10.1016/j.jtice.2020.10.008

[68] Tiwari BK. Ultrasound: A clean, green extraction technology. TrAC Trends in Analytical Chemistry, 2015;71: 100–109. DOI: https://doi.org/10.1016/j.trac.2015.04.013

[69] Babu SG, Neppolian B, Ashokkumar M. (2016).Ultrasound-Assisted Synthesis of Nanoparticles for Energy and Environmental Applications. In: Handbook of Ultrasonics and Sonochemistry. Springer, Singapore. DOI: https://doi.org/10.1007/978-981-287-278-4_16

[70] Baig RB, Varma RS, Alternative energy input: mechanochemical, microwave and ultrasound-assisted organic synthesis. Chem. Soc. Rev., 2012; 41: 1559–1584. DOI: https://doi.org/10.1039/C1CS15204A

[71] Murcia MJ, Shaw DL, Woodruff H, Naumann CA, Young BA, Long EC. Facile Sonochemical Synthesis of Highly Luminescent ZnS-Shelled CdSe Quantum Dots. Chem. Mater., 2006; 18: 2219–2225. DOI: https://doi.org/10.1021/cm0505547

[72] Jameel MS, Aziz AA, Dheyab MA. Impacts of various solvents in ultrasonic irradiation and green synthesis of the platinum nanoparticle. Inorg. Chem. Commun. 2021;128: 108565. DOI: https://doi.org/10.1016/j.inoche.2021.108565

[73] Wang J, Nie P, Ding B, Dong S, Hao X, Dou H, Zhang X. Biomass derived carbon for energy storage devices. J. Mater. Chem. A. 2017; 5(6): 2411–2428. DOI: https://doi.org/10.1039/c6ta08742f

[74] He X, Li R, Qiu J, Xie K, Ling P, Yu M, Zhang X, Zheng M. Synthesis of mesoporous carbons for supercapacitors from coal tar pitch by coupling microwave-assisted KOH activation with a MgO template. Carbon, (2012; 50(13): 4911–4921. DOI: https://doi.org/10.1016/j.carbon.2012.06.020

[75] Biswal M, Banerjee A, Deo M, Ogale S. From dead leaves to high energy density supercapacitors. Energy Environ. Sci. 2013; 6(4): 1249. DOI: https://doi.org/10.1039/c3ee22325f

[76] Zhao X, Johnston C, Grant PS. A novel hybrid supercapacitor with a carbon nanotube cathode and an iron oxide/carbon nanotube composite anode. J. Mater. Chem. 2009; 19: 8755-8760. DOI: https://doi.org/10.1039/B909779A

[77] Raymundo-Piñero, E., Leroux, F., & Béguin, F. (2006). A High-Performance Carbon for Supercapacitors Obtained by Carbonization of a Seaweed Biopolymer. Advanced Materials, 18(14), 1877–1882. DOI: https://doi.org/10.1002/adma.200501905

[78] Zhu C, Akiyama T. Cotton derived porous carbon via a MgO template method for high performance lithium ion battery anodes. Green Chemistry. 2016; 18(7): 2106–2114. DOI: https://doi.org/10.1039/c5gc02397a

[79] Li X, Xing W, Zhuo S, Zhou J, Li F, Qiao S, Lu G, Preparation of capacitor's electrode from sunflower seed shell. Bioresour. Technol. 2011; 102: 1118-1123. DOI: https://doi.org/10.1016/j.biortech.2010.08.110

[80] Subramanian V, Luo C, Stephan AM, Nahm KS, Thomas S, Wei B. Supercapacitors from Activated Carbon Derived from Banana Fibers. J. Phys. Chem. C, 2007; 111: 7527-7531. DOI: https://doi.org/10.1021/jp067009t

[81] Guo Y, Qi J, Jiang Y, Yang S, Wang Z, Xu H. Performance of electrical double layer capacitors with porous carbons derived from rice husk. Mater. Chem. Phys., 2003; 80: 704-709. DOI: https://doi.org/10.1016/S0254-0584(03)00105-6

[82] Xing W, Xue J, Dahn J. Optimizing Pyrolysis of Sugar Carbons for Use as Anode Materials in Lithium-Ion Batteries. J. Electrochem. Soc. 1996; 143: 3046-3052. DOI: https://doi.org/10.1149/1.1837162

[83] Kali R, Padya B, Rao TN, Jain PK. Solid waste-derived carbon as anode for high performance lithium-ion batteries. Diam. Relat. Mater. 2019;107517. DOI: https://doi.org/10.1016/j.diamond.2019.107517

[84] Tao L, Huang Y, Zheng Y, Yang X, Liu C, Di M, Larpkiattaworne S, Nimlos MR, Zheng, Z. Porous carbon nanofiber derived from waste biomass as anode material in lithium-ion batteries. J. Taiwan Inst. Chem. Eng. 2019; 95: 217-226. DOI: https://doi.org/10.1016/j.jtice.2018.07.005

[85] Zhu C, Akiyama T. (2016). Cotton derived porous carbon via a MgO template method for high performance lithium ion battery anodes. Green Chemistry, 2016;18(7): 2106–2114. DOI: https://doi.org/10.1039/c5gc02397a

3

Production of Energy from Biowaste: An Overview of the Underlying Biological Technologies

Brij Bhushan and Arunima Nayak

Department of Chemistry, Graphic Era University, Dehradun, Uttarakhand, India
E-mail: brij.chem@geu.ac.in

Abstract

The demand for bioenergy is fast picking up with growing climate change policies and with a growing need for independence from fossil-based energy. Biowaste has been identified as a low-cost, widely available renewable source for the sustainable generation of biofuels. Various advantages have been cited for use of biowaste for biofuel generation like reduced dependency on fossil reserves for energy, minimized greenhouse gas emissions, minimized health and environmental risks, and reduced environmental burden of biowaste disposal. The liquid biofuels like bioethanol and biodiesel generated via microbial fermentation and transesterification are widely in use as a transportation fuel and steadily replacing traditional fuels like gasoline and diesel. Gaseous biofuels like biogas are upgraded to generate electricity for household purposes, used as transportation fuel, besides being used for cooking or lighting purposes. Global production of biofuels in the year 2018 was 154 billion liters and the forecast is that the production rate is expected to increase to 25% by 2024. The major use of biofuels is expected to be in the transport sector. Since transport accounts for 30% of global energy usage, the use of biofuels is important for ensuring a sustainable energy future. With this backdrop, the chapter gives an overview of the different biological technological transformations of biowastes to biofuels (biogas, bioethanol, biodiesel, and biohydrogen). The advancements made to date in anaerobic digestion, microbial fermentation, and dark fermentation are discussed, and the different biowastes used successfully are highlighted along with critically discussing of the major hurdles that have led to lower biofuel yield. The future scope rests on strong policy support along with innovation to reduce costs. This will ensure the scaleup of biofuel consumption and application.

Keywords: Biofuels, Sustainability, Biogas, Bioethanol, Biodiesel, Biohydrogen.

3.1 Introduction

The demand for energy is increasing over the last decade because of increasing population growth, increased industrialization, and increasing consumerism/lifestyle changes. Since the year 2010, it is estimated that the rate of consumption of energy will nearly double by the year 2050 from 14TW to 28TW [1]. The majority of the energy demand (approximately 80%) is being met by fossil fuel based natural resources (coal, oil, natural gas, etc.) which are not only getting depleted but also are the major contributors to greenhouse gas emissions and climate change. Thus, for maintaining a sustainable energy demand-supply chain and for ensuring the security of the energy resources, more emphasis is being made by researchers, scientists, technocrats, etc. on renewable energy sources which are abundantly available, are less costly, and are environmentally friendly. In this regard, bio-waste is a potential renewable and sustainable source for energy production. As per the revised EU (European Union) Waste Framework Directive7 [2], bio-wastes are biological in origin and are categorized as (a) garden and park waste and (b) food and kitchen wastes originating from households, restaurants, and other food processing plants. Studies have shown that irrespective of the country of origin, the generation of food waste is high and is the biggest contributor to bio-wastes. As per a survey conducted by EPA (Environment Protection Agency) [3], out of the total generated MSW (municipal solid waste) of 293 million tons in 2018, 35% was represented by bio-wastes. Food waste was the fourth largest contributor accounting for approximately 63 million tons or 22% of total MSW generated in 2018. As per reports by FAO (Food and Agricultural Organization) of the United Nations, approximately one-third of the food production led to food waste, and approximately 1.3 billion tons of food waste are discarded annually and is expected to increase to 2.2 billion tons by 2025 [4]. The major reason cited for the large-scale food wastage is economic growth and modern lifestyle changes. Such wastes are characterized by high rates of biodegradation. The recycling methods usually adopted by local communities are composting and animal feed while the major disposal methods are landfilling and incineration [5]. Such disposal leads to not only soil and water pollution, but the alarming rate of emission of methane gas is 23 times more damaging to the environment and contributor to climate change as compared to CO2 as a greenhouse gas [5]. The alarming rate of generation of such biological wastes is not only a challenge to the urbanites but the unscientific methods adopted for the disposal have added risk to the environment, economy, and society. Under the smart waste management options and as per the new Waste Directive 2008/98/EU [2] recycling or the bio-refinery of the bio-waste to produce commercial by-products is fast gaining momentum in the scientific community as a sustainable option. In this aspect, recovery of energy from bio-waste is an attractive alternative to non-renewable energy sources. Firstly, this helps in reducing the dependence on fossil fuel reserves for meeting the energy demand, and consequently, the depletion of natural resources is reduced. Secondly, the use of bio-waste to meet the energy demands of society helps to minimize the risk to human health and the environment via fewer greenhouse gas emissions, thereby an overall balance in the ecosystem is maintained. Thirdly the bio-refinery concept helps to reduce the environmental burden of disposal of bio-waste.

Various reviews have addressed the different valorization technologies for the recovery of energy from food wastes and other bio-based wastes [6–9]. With this background, the outline of the chapter is to address the different technologies used for the recovery of biofuels like biogas, bio alcohol, biohydrogen, bio oil, etc. from biowaste. While the bio-based conversion technologies involve the use of anaerobic as well as microbial fermentation to yield bio-methane, bio-alcohol, and bio-hydrogen; the thermal process involves incineration, hydrothermal liquefaction, and carbonization of the biowaste for generating bio-oil/syngas. Besides addressing an up-to-date advancement, advantages, and limitations of each technology, the hurdles or challenges faced during the treatment of different biowastes are discussed. Finally, future perspectives have been proposed for more effective treatment of biowaste for the generation of renewable energy.

3.2 Current Technologies for Energy Generation from Biowaste

The bio-refinery or valorization of biowaste to generate energy has emerged as a sustainable pathway for the unscientific management of such wastes and on account of the rising costs of energy due to depleting fossil reserves. Unscientific disposal methods adopted for biowastes, as well as the use of fossil fuel-based resources for energy generation, result in environmental pollution and degradation. The favorable outcomes because of the bio-refinery pathway are not only connected with clean alternative energy generation, environmental protection, protection of natural resources, and waste management but also significant revenue generation. Also, the biofuels generated from biowastes do not lead to food versus fuel competition and hence are sustainable. Scientific studies have demonstrated favorable outcomes with respect to biofuel generation from various types of wastes of biological origin. The major sources of biowaste generation are food processing, agro-food industrial processing, animals, hotels, municipalities, etc. Irrespective of their source, the major characteristics of the biowastes are the high moisture content. Also, plant or agriculturally based wastes (for e.g., straw and stalk of cereals, sugarcane bagasse, peels of fruits, pomace of grape, apple, etc.) have a high amount of lignocellulose, carbohydrates, and sugars, polyphenols, etc. All of which require appropriate pre-treatments prior to the energy conversion processes. But despite this, such biowastes have demonstrated good yields of biofuels.

The major technologies adopted for biofuel generation are anaerobic digestion, transesterification, and microbial fermentation (all of which involve biotechnological transformations). Biological processes require either enzymes or microbial species for a breakdown of lignocellulosic components. The high moisture content does not inhibit the biological transformation of biowastes. Also, the process involves maximum biofuel yield accompanied by minimum emissions; hence is more acceptable for industrial applications. But the major demerit is the lengthy time required to bring about enzymatic or microbial biotransformation. Incineration involves thermal processes and undergoes physico-chemical transformations of the biowaste. This thermal process of biowaste management involves not only a high-temperature regime but also is associated with gaseous emissions (CO2, CH4, NH3, HCN, CO, H2) and hence is not environmentally friendly. The high moisture content is a major deterring factor that affects the efficiency of the thermal process adversely. A detailed insight into the various technologies, along with their pros and cons with respect to the efficient utilization of biowaste and the corresponding yield of biofuels are described in the following sub-sections.

3.3 Anaerobic Digestion for Generation of Biogas

Anaerobic digestion is the technology of stepwise degradation of the different organic components in biowastes under controlled operational conditions in a closed vessel (bioreactor) via a series of bio-based chemical reactions to yield biogas whose composition is methane (50–70%), carbon dioxide (20–50%), hydrogen and other gases (in trace amounts) [10]. Methane content of 55–75% in biogas implies an energy content of 6–6.5 kWh/m3. In addition to biogas, a solid fraction is generated which has the potential to be used as a fertilizer. As per studies conducted by Murphy et al [11], it has been estimated that if the anaerobic digestion process is operating at an efficiency of 35% then 1m3 of biogas generated is equivalent to 21 MJ of energy, and it has the capacity to generate 2.04 kW h of electricity. Some of the major uses of biogas are (a) for heating purposes for maintenance of equipment (b) for domestic cooking and lighting purposes (c) for heating in the beverage industry and ethanol production (d) for natural gas supply for domestic purposes via injection into the grid (e) for generation of electric power (f) as a fuel for civil transport and road vehicles (g) use in fuel cells for generation of electricity. Thus, to achieve higher biogas yields, an insight into the technological background of the anaerobic digestion process and factors affecting the stable performance of the process is required.

The stepwise degradation process via a series of chemical steps (hydrolysis, acidogenesis, acetogenesis, and methanogenesis) takes place by the action of the individual class of bacteria which have different characteristics, different operational conditions, and different functions [12–14]. The starting of the four steps is hydrolysis (catalyzed by hydrolytic bacteria) which involves the breakdown of the macromolecules in biowaste (carbohydrates, lipids, and proteins) to form simple sugars, fatty acids, and amino acids respectively. The next step of degradation is acidogenesis which under the action of acidogenic bacteria produces volatile organic acids besides carbon dioxide, hydrogen, ammonia, and alcohol. Acetogenic bacteria which catalyze the degradation of acetic acid, hydrogen, and carbon dioxide are much slower in their activity as compared to acidogenic bacteria. The last step which involves the formation of methane, carbon dioxide, and hydrogen is catalyzed by two separate classes of bacteria. Slow-growing acetoclastic methanogens are responsible for producing methane from acetic acid while the hydrogenotrophic methanogens utilize hydrogen and carbon dioxide to produce methane.

Higher yields along with a better quality of biogas (having a higher fraction of methane) depend on operational and environmental parameters (like strict anaerobic conditions, control of temperature, pH, C:N ratio, and redox potential), physico-chemical characteristics of biowaste and reactor configuration. Strict maintenance of the operational parameters during the digestion process helps to establish optimum conditions for the healthy growth and metabolism of the bacterial community; thereby establishing a synergistic relationship between the different classes of bacteria. Optimum temperature ranges of 35–40 °C (mesophilic) and 50–65 °C (thermophilic) while a neutral pH range of 6.3–7.8 is found to be optimum for a stable digestion process.

Besides the operational factors, the physical and chemical characteristics of the waste matrix-like the presence of moisture, volatile solids, carbon, nitrogen content, etc. also play a decisive role in the quality and yield of biogas [15]. The carbon-nitrogen ratio of 25-30:1 has been identified as crucial for the stability of the digester operation. While low nitrogen content in the biowastes can lead to a slowdown of the digestion process, higher nitrogen content can result in an increased accumulation of ammonia. Both conditions are not suitable for the smooth working of the reactor. Again, high lipid or high salt concentration in biowaste can have an inhibitory effect on the digestion process and hence on the yield of biogas. The addition of two or more substrates during the digestion process is known to enhance the biogas yields via either incorporating some missing nutrients or may be due to a positive synergistic effect [16]. In one such study, enhanced biogas yield was obtained due to the addition of sewage sludge as a co-digestate to the reactor containing food waste as the substrate [17]. The addition of abattoir wastewater to the anaerobic digestion of fruit and vegetable waste helped to enhance the biogas yield to 51.5% [18]. Similarly, it was reported that co-digestion of food waste with cow manure carried out in an anaerobic mesophilic reactor helped to enhance the bio-methane yield by 26% as compared to the individual yields obtained from the digestion of food waste and manure [19]. For certain biowastes having complex structures, the digestion process suffers from longer solid residence time and slower conversion efficiency. Research studies have shown that an initial pre-treatment of the biowaste helps to accelerate the digestion process causing enhanced biogas yield [20–22]. The different pre-treatments undertaken so far are classified as chemical, physical and biological methods. Among the chemicals used, ozonation of organic waste has resulted in 37% enhanced biogas yield [23]. Thermal [24] and mechanical [25] methods are two physical pre-treatments that have brought about enhanced hydrolysis and hence higher biogas yield.

For hard to degrade biowastes, the use of microwave [26] and ultrasound [27] has brought about better and faster solubilization at a lesser energy requirement; thereby enhancing the efficiency of the digester performance for higher biogas yields as compared to the use of thermal pre-treatment methods. Tiehm et al. demonstrated a reduced retention time of 12 days during the anaerobic digestion of sewage sludge using ultrasonication as a pre-treatment method [27]. Biological pre-treatment methods like adding different microbes like fungi, mature compost, or enzymes to the anaerobic digester have initiated higher solubilization of the biowaste and hence better biogas yields [28]. Although such pre-treatments are environmentally and energetically favorable, the hydrolysis rate is much longer as compared to physical and chemical pre-treatments.

The reactor in which the digestion process takes place to yield biogas must have an appropriate design and configuration to allow the healthy growth of the microbial community and to allow maximum contact between the microbial community and the biowaste [29, 30]. Studies have further confirmed the correlation between reactor design/configuration with high biogas yields. As far as the reactor configuration is concerned, studies have shown that a two-stage reactor separating the hydrolysis/acidogenesis and acetogenesis/methanogenesis is more favored as compared to a single-stage reactor [31]. This is mainly on account of different operational factors that need to be maintained during the hydrolysis and methanogenesis for effective growth and metabolism of the respective class of bacteria; thus, a two-stage reactor allows the different bacteria to function better leading to less pH inhibition issues that are frequently encountered in single-stage reactors. Various reactors have been designed and are operational. The conventional reactors in use are the anaerobic sequencing batch reactor (ASBR), continuous stirred tank reactor (CSTR), and anaerobic plug flow reactor (APFR) [32]. A single-stage reactor is operational in both ASBR and CSTR. ASBR has the capacity to operate under a wide range of influent volumes, but a CSTR can handle a fixed influent flow rate [33]. A major criterion during the operation of a bioreactor is to ensure biomass retention. In both the reactors (ASBR and CSTR), due to their inability to retain biomass, the maximum of the microbes is washed out along with the effluent. Another disadvantage is the accumulation of fatty acid during a CSTR operation which results in inhibition and hence less biogas yield. The design aspect of AFPR is superior as compared to the CSTR in the sense that it ensures high biomass retention along with less inhibition and hence has offered a stable operation. The high-rate anaerobic reactors like the anaerobic contact reactor (ACR), up-flow anaerobic sludge bed reactor (UASB), up-flow anaerobic solid-state reactor (UASS), and anaerobic baffled reactor (ABR) exhibit not only enhanced biomass retention but also, allow a dense growth of microorganisms [34, 35]. The solid retention time is much higher compared to the hydraulic retention time.

Table-3.1 has summarized the studies carried out on anaerobic digestion of different biowastes with corresponding biogas yields achieved. It is observed that there is a variation in the biogas yields depending on the experimental conditions under operation during the digestion process. As per a study conducted via a full-scale operational digester [43], it was found that the total energy recovered during anaerobic digestion of biowaste consisting mainly of food waste was 1462 kW h per ton of total solid TS in the food waste; biogas yield was 642 m3/ton of volatile solid added. The methane content in the biogas was 62%. Anaerobic digestion center in Toyohashi, Japan (operational since 2017) utilizes 59 tons/day of biowaste and 472 m3/day of sludge as co-digestate to generate 6,800,000 kWh/year of power sales volume which is available to approximately 1,890 families (with the assumption that each household utilizes 300 kWh per month) [47]. Such an optimistic outcome suggests the potential of biowaste as an alternative substrate for energy generation. The anaerobic digestion technology has the capacity to bring about complete degradation of biowaste in a clean process but also it can generate biogas in a cost-effective manner.

3.4 Microbial Fermentation for Bioethanol Generation

Among the different bio-alcohols of commercial interest that are generated from microbial fermentation of the biowastes are bioethanol and biobutanol. Both bioethanol and biobutanol are used as alternative transportation fuels for gasoline. They are also used as additives to gasoline in spark-ignition engines mainly because of their higher-octane number and reduced gaseous emissions. Biobutanol has superior physicochemical properties (lower vapor pressure, improved combustion efficiency, and higher energy density) that can enable enhanced engine performance but the biotechnology for its production is complicated [48].

Bioethanol generated from food wastes/agricultural wastes/municipal solid wastes/agro-processed wastes etc. has various potential benefits like less dependency on petroleum and gasoline leading to less greenhouse gas emissions along with the reduced cost of production, better waste management, and reduced cost of waste disposal. Also, bioethanol generation is

Table 3.1 Biogas yield performance from anaerobic digestion of biowaste.

Biowaste	Co-digestion	Bioreactor	Duration (days)	Biogas yield	CH_4 yield (% CH_4)	References
Cattle slaughterhouse waste	Nil	1000ml serum bottle	30	–	641L/kg VS	[36]
Potato waste	Nil	CSTR (0.5L)	60 days HRT 20days	–	0.42m^3/kg (62%)	[37]
Potato waste	Beet leaves	CSTR (0.5L)	60 days HRT 20days	–	0.68m^3/kg (84%)	[37]
Food waste	Nil	3 stage bioreactors	NA HRT: 12days	223L/kg sCOD degraded	NA	[38]
Potato solid waste & activated sludge	Nil	2 stage lab scale UASB reactor (2L)	50	NA	0.39m^3/kg	[39]
Cafeteria food waste	Nil	3 stage CSTR reactors (0.05m^3 and 2m^3)	–	391 L/kg COD 373 L/kg COD	280 L/Kg COD (72%) 254 L/kg COD (68%)	[40]
Food waste (University mess)	Cattle manure	1L glass digester	–	–	388 ml/g VS/day (62%)	[41]
Food waste (University mess)	Nil	1L glass digester	–	–	347 ml/g VS/day (61%)	[41]
Fruit and vegetable waste	Cattle and swine slaughterhouse waste	Semi-continuous (2L)	HRT 50days	817ml/day	0.3 m^3/kg VS (54%)	[42]
Kitchen and garden waste	Nil	2 stage bioreactors	200	642m^3/ton VS	402m^3/ton VS (62%)	[43]
Coffee waste	Sewage sludge	160ml serum vials	144 days		0.28m^3/kg VS (85%)	[44]
Food waste	Sewage sludge	13L CSTR and 2L Anaerobic membrane bioreactor	–	0.893 L/g VS	59.2%	[45]
Food waste	Tylosin fermentation dreg	NA	–	–	399ml/g VS	[46]
Food waste	Nil	NA	–	–	256 ml/g VS	[46]

environmentally friendly and sustainable. Various biowastes like food waste from the household, canteen, cafeteria, restaurants, etc., agricultural wastes like coffee wastes, banana peels, rice straw, wheat straw, sugarcane bagasse, grape pomace, apple pomace, sugar beet pulp, pineapple waste, potato peel waste, etc. have shown great potential in bioethanol generation. A summary of their performance is outlined in Table-3.2.

As evident from Table-3.2, the technology used for ethanol generation is initial saccharification or hydrolysis which involves enzymatic degradation of

Table 3.2 Performance of microbial fermentation for bioethanol generation from biowastes

Food Waste	Technology	Hydrolysis	Microorganism	Yield of ethanol	Reference
Coffee pulp	SHF	Distilled water (4hrs)	*S. Cerevisiae*	7.4g/L	[49]
Kinnow waste Banana peel (Steam explosion pretreatment)	SSF	enzyme	*Saccharomyces cerevisiae G and Pachysolen tannophilus*	0.426	[50]
Food waste	SSF	enzyme	*S. Cerevisiae*	57.6g/L	[51]
Food waste from University Cafeteria	SSF	enzyme	*S. cerevisiae and Zymomonas mobilis*	0.36g/g of FW	[52]
Rice hull	SSF	Lime pre-treatment	*S. Cerevisiae*	154mg/g	[53]
Food waste from University Cafeteria	SSF	enzyme	*S. Cerevisiae*	0.31g/g TS	[54]
Food waste from University Cafeteria	SHF	enzyme	*S. Cerevisiae*	0.43g/g TS	[54]
Ripe Banana peels	SSF	enzyme	*S. Cerevisiae*	28.2 g/L (0.29g/g)	[55]
Kinnow wastes (peel and pulp)	SSF	enzyme	*S. Cerevisiae*	42g/L	[56]
Pie waste	SSF	enzyme	*S. Cerevisiae*	0.329g/g	[57]
Food waste from University Cafeteria	SSF in 160ml serum bottle	enzyme	*G. thermoglucosidasius (ATCC 43742) and T. ethanolicus (ATCC 31938)*	13g/L	[58]
Barley straw Corn stalk	SHF	enzyme	*Saccharomyces cerevisiae and Pachysolen tannophilus ATCC 32691*	10g/L 5g/L	[59]
Noodle waste	SSF	enzyme	*S. Cerevisiae*	60.7 g/L	[60]
Sugarcane bagasse	SHF	HCl	*Candida shehatae NCIM 3501*	0.48g/g	[61]
Rice straw	SHF	enzyme	*S. Cerevisiae*	0.096 g/g (25.56 g/L)	[62]
Corn stover	SSF	enzyme	*S. Cerevisiae*	59.8 g/L	[63]
Corn stover	SHF	enzyme	*S. Cerevisiae*	50.4 g/L	[63]

complex carbohydrates to simple reducing sugars like hexose and pentoses followed by fermentation via yeasts like S. cerevisiae. This technology has been performed either via two separate steps of saccharification and fermentation (separate hydrolysis and fermentation-SHF) or via a single step involving both saccharification and fermentation (sequential saccharification and fermentation-SSF). The saccharification step is catalyzed by the action of enzyme cellulase which has the highest operational efficiency at a temperature of 45 °C and pH of 4.8 in 50mM Na-acetate buffer. Whereas the conditions required for optimum operation of S. cerevisiae for ethanol generation yield during the fermentation step are 37 °C and a pH of 5 in a 100mM of phosphate buffer. Various advantages have been demonstrated from the SSF technology like less inhibitory action of glucose and cellobiose on cellulase activity and hence higher ethanol yield as well as less processing time. Also, fewer costs are incurred via the use of only one bioreactor under the SSF technology. E.g., in a study conducted on wheat straw [64], it was found that SSF technology brought about 68% of theoretical ethanol yield (per unit of biomass) as compared to 81% under the SHF technology. But because the overall time duration under SSF (30 hrs) was less compared to the same under SHF (96 hrs), the ethanol productivity (grams per unit volume per unit time) was much higher in the SSF.

Because of the hard to digest properties of certain biowastes like those having lignocellulosic components, it is imperative to undertake pre-treatment methods like mechanical methods, extrusion, liquid hot water bath, steam explosion, ammonia fiber expansion, use of acid, alkali, heat, and enzymes for enhancing saccharification/hydrolysis of the carbohydrate fraction. While the use of acid, alkali, and enzymes has led to efficient hydrolysis of lignin and cellulose components leading to enhanced glucose yields, mechanical methods, extrusion, liquid hot water baths, and steam explosion are not cost-effective and energy-efficient. Similarly, parameters like temperature, pH, glucose levels, presence of organic acids, and immobilization of yeast are known to affect the ethanol production rate. Thus, effective pre-treatment steps along with optimization of parameters affecting the yield of glucose and ethanol are essential for improving the efficiency of the technology. Also, as evident from Table-2, the majority of studies have been conducted using S. cerevisiae as the microbial strain for bringing about fermentation to ethanol. But S. cerevisiae has the capacity to ferment only hexoses thereby bringing about lower ethanol yields. Ban-Koffi and Han [65] have used Zymomonasmobilis for fermentation of pineapple waste while Korkie et al., [66] have used Pichia rhodanese onto grape pomace for ethanol generation. It was found that both species have the potential for utilizing pentose sugars too for conversion to ethanol.

3.5 Microbial Fermentation for Bio-Hydrogen Generation

Hydrogen derived from conventional energy sources has an energy content of 120MJ/kg as compared to other fossil fuel sources [67–69]. Approximately 90% of hydrogen is recovered from fossil reserves. The current methods for hydrogen generation (pyrolysis, hydrocarbon reforming, partial oxidation of fossil fuels, plasma technology, coal gasification) are energy-intensive and are major emitters of carbon dioxide; hence are not environmentally friendly. Based on these facts, hydrogen derived from renewable sources like biowaste has great potential for use as an alternate energy source. It is clean, is not a greenhouse gas emitter, and has high energy content. It can be easily converted to heat or electricity; can be used as a fuel for use in rockets, automobiles, locomotives, and other transportation modes. The attractive feature of the use of hydrogen as a transportation fuel is that it has zero emission and high efficiency. More importantly, hydrogen is abundantly available. Biological methods offer great opportunities for generating hydrogen from biowastes via cost-effective and eco-friendly technologies. Biological methods involve microorganisms (bacteria and algae) to initiate the fermentation of various carbon-based wastes for sustainable hydrogen production. Such methods thus not only recycle the biowastes to generate commercially important biofuels like hydrogen but also, play a major role in the safe and clean disposal of organic wastes.

The anaerobic dark fermentation technology used for hydrogen generation involves the initial generation of reducing sugars via enzymatic (cellulase) hydrolysis of the carbohydrate component of biowaste followed by thermophilic heterotrophic fermentation of the sugars to yield organic acids. Finally, hydrogen is generated from the organic acids via the photo heterotrophic fermentation process. The microorganisms that have shown great potential for hydrogen generation are anaerobic bacterial species like Clostridium, facultative bacterial species like Enterobacter, and thermophilic bacteria like Thermotoga. Such bacterial species possess high growth rates and can thrive on various carbohydrate-rich substrates including biowastes and wastewaters. The biowastes that have shown good hydrogen production efficiency are food wastes from households, restaurants, cafeterias, agro-food industrial wastes, etc. Table-3.3 summarizes the performance of dark fermentation for bio-hydrogen generation from biowastes

Table 3.3 Performance of dark fermentation for bio-hydrogen generation from biowastes

BioWaste	Microorganism	Reactor (temp, pH)	Yield of bio-hydrogen	Reference
Apple processing wastewater (9 g-COD/L)	Mixed culture	Batch-250mL bottles	0.9 L H_2/L medium (0.1 L H_2/g COD)	[70]
Potato processing Wastewater (21 g COD/L)	Mixed culture	Batch-250mL bottles	2.1 L H_2/L medium (0.1 L H_2/g COD)	[70]
Candy wastewater (20 g COD/L)	Mixed culture	Batch-250mL bottles	2 L H_2/L medium (0.1 L H_2/g COD)	[70]
Food waste from a dining hall	*Clostridium sp*	Continuous: 3.8 L anaerobic leaching bed reactor	0.39 L H_2/g COD	[71]
Food waste from residential home	Mixed culture (anaerobic sludge)	Batch: 1 L glass bottle (50 °C)	57 ml H_2/g VS	[72]
Rice slurry (5 g CHO/L)	Mixed culture	Batch 280 mL serum bottles (pH 4.5; 37 °C)	346 ml H_2/g carbohydrate (62%)	[73]
Starch wastewater	*Thermoanaero-bacterium sp* mixed culture	Batch: 280 mL glass bottles (pH 6; 55 °C)	92 ml H2/g starch (60%)	[74]
POME	Mixed culture	ASBR (3.2L)	6.33 L H_2/L-POME	[75]
Cheese whey (49.2 g lactose/L)	*C. saccharoper-butylaceton-icum ATCC 27021*	Batch: 250 mL glass bottle pH 6	7.89 mmol H_2/g lactose	[76]
Dairy waste (10.4 g COD/L)	Mixed culture	AnSBR (100 °C; pH 6)	1.105 mmol H2/m3/min	[77]
Food waste from the school cafeteria	Mixed culture obtained from an anaerobic digester	ASBR	0.54 mol H_2/mol hexose added (44%)	[78]
Food waste from the school cafeteria (pretreatment with alkaline shock treatment	Mixed culture obtained from an anaerobic digester	ASBR	0.90 mol H_2/mol hexose added (44%)	[78]

Some of the important factors that have been identified to be directly related to hydrogen yield are the composition of biowaste, nature of the inocula, product inhibition as well as operational parameters like pH, and temperature [79]. Various studies have strongly recommended the higher yield of hydrogen using carbohydrate-rich biowastes as compared to protein-rich or fat-rich wastes (Table-3.3). The optimum pH that has resulted in the highest hydrogen production is in the range of 5–7. Operation under thermophilic conditions has resulted in higher hydrogen generation as compared to operation under mesophilic conditions. For continuous operations, hydraulic retention time is an important parameter affecting hydrogen production. Volatile fatty acids, ethanol generated as a by-product during the final hydrogen generation as well as excessive glucose formed during the enzymatic hydrolysis are the major inhibitors of the process efficiency. To regulate the inhibiting action and cause enhanced hydrogen yield, pretreatment of biowastes has resulted in the suppression of lactic acid bacteria which are the major source of inhibitors. The major pretreatments employed as per various studies are heating, acid/alkaline hydrolysis, and ultrasound-assisted hydrolysis [80].

To overcome the limitations of dark fermentation technology under both batch and continuous operations (for e.g., reduced hydrogen yield and high production cost leading to questionable large scale hydrogen production), studies have been proposed for a combined dark and photo fermentation technology for ensuring increased hydrogen generation from biowaste [81]. In this technique, the lactic acid produced because of hydrolysis of biowaste (a major inhibitor) is converted to hydrogen by the action of photo fermentative non-sulfur bacteria. A two-stage dark fermentation and anaerobic digestion have been proposed for both hydrogen and methane generation from biowastes [82]. The advantages reported for use of dual technology are higher hydrogen yields, better process efficiency, and overall reduced cost.

3.6 Transesterification for Biodiesel Generation

Biodiesel is a biofuel recovered from vegetable-based oils, animal fat, or microbial oil (generated from the fermentation of biowaste) [83]. It is mainly used as a transportation fuel. It provides 93% more usable energy as compared to fossil-based energy. Also, its use is associated with 41% less greenhouse gas emissions as compared to the use of diesel; reduces toxic emissions, and has a minimal adverse impact on the environment. Besides the environmental aspect, the use of biodiesel has been found to improve the lubricity and cetane number of fuels. Biodiesel is used directly in the pure form as a fuel (B100) or is mixed in different proportions to produce different grades of fuel (for e.g., B20 fuel has 20% biodiesel and 80% diesel or B5 has 5% biodiesel and 95% diesel) [84].

The technology involves the initial extraction of oil/lipids etc. from different biowastes via conventional extraction, Soxhlet extraction, and supercritical fluid extraction [85, 86]. During conventional extraction, complete recovery of oil or lipids from biowaste is difficult and the process involves the use of excess volatile organic solvents. On the other hand, during Soxhlet extraction, complete recovery can be expected along with no requirement for excess volatile organic solvents. Supercritical fluid extraction involves green solvents, and the process is sustainable ensuring complete recovery of the extracted oil/lipids. The transesterification process involving the use of chemicals, or enzymes as catalysts to yield biodiesel. The use of chemicals as catalysts requires less reaction time and reduced costs as compared to the use of enzymes as catalysts. Among the chemicals used, alkali catalyzed transesterification yields high-quality biodiesel in a short reaction time, but the process is not feasible for wastes having high free fatty acid components [87]. A two-step transesterification involving initial acid-catalyzed esterification followed by alkali catalyzed transesterification was used for handling high free fatty acid-based biowastes. But the disadvantages are the long reaction time and low recovery rate of the catalysts. More recently, the microwave-assisted catalytic transesterification process has yielded a high yield as well as high-quality biodiesel [88, 89]. Such a process has been efficiently tested on oil extracted from different biowastes like rapeseed, cotton seed, and waste cooking oils [90–92]. The performance of the transesterification process for biodiesel generation from oil/lipid extracted from biowaste is summarized in Table 3.4.

3.7 Discussion on Potential Challenges and Solutions for Biofuel Generation and Application

Biowastes are characterized by variability in the volume, time, and site of generation, as well as variability in composition. Because of their biological nature, mass scale landfilling and incineration has resulted in greenhouse gas emissions, global warming, and climate change. All of which are unwarranted to the environment. Technical and sustainable methods of such waste management have paved the pathway for their biorefinery to yield biofuels. Against the backdrop of excessive and ever-increasing demand for energy along with limited fossil fuel reserves, biofuels generated from biowastes

Table 3.4 Performance of transesterification process for biodiesel yield from different biowastes

Biowaste	Biodiesel yield	Process description	Reference
Noodle waste oil	98.5%	Transesterification using methanol and KOH as catalysts (methanol: oil 1:8) at 60 °C for 2hrs	[60]
Noodle waste oil	97.8%	Transesterification using methanol and H_2SO_4 as catalysts (methanol: oil 1:6) at 80 °C for 3hrs	[60]
Waste cooking oil	92%	pretreatment with sulfuric acid to remove FFAs and a microwave transesterification process in the presence of KOH as a catalyst	[93]
Waste cooking oil	93%	Transesterification using methanol and alkali (KOH) as a catalyst	[94]
Waste cooking oil	95.9%	Transesterification using methanol and CaO (extracted from pyrolysis of snail shell) as catalyst (methanol: oil- 6:9)	[95]
Duck tallow	97%	Transesterification with methanol in the presence of KOH at 65 °C	[96]
Beef tallow	76.8	Transesterification with methanol in the presence of KOH	[97]
Chicken tallow	91.5%	Transesterification with methanol in the presence of KOH	[97]
Pork lard	90.8%	Transesterification with methanol in the presence of KOH	[97]
Chicken and swine fat residue	83%	Transesterification with methanol in the presence of KOH at 30 °C	[98]
Fish waste	96%	Transesterification with methanol in the presence of β-tricalcium phosphate (extracted from calcination of fish bones) (methanol: fish oil-6.5:1) at 30 °C	[99]
Waste sardine oil	94%	Transesterification with methanol in the presence of lipase *Aspergillus niger* enzyme supported on activated carbon as catalyst (methanol: oil-9:1) at 30 °C	[100]

will be the major source of energy in the coming decades and will be the major drivers of economic growth for many countries. Based on various studies on the economics and environmental implications of different biofuel generation on an academic level, it is revealed that biofuels can have the potential to lower greenhouse gas emissions. Also, the production of biofuels from biowastes can lower the dependency both on petroleum products and on petroleum-rich countries. This will ensure lower the cost of fossil fuels and will make the countries economically self-reliant. Liquid biofuels like biodiesel and bioethanol are used as transportation fuels and are slowly but steadily replacing diesel- and gasoline-based fuels. In high-income countries, biogas is primarily used to generate electricity and heat whereas, in low-income countries, the major use of biogas is for cooking and lighting purposes. Countries like Germany, Italy, France, Switzerland, etc., have started biogas production on a commercial scale. The major use of biohydrogen is in the automotive sector and in the production of nitrogenous fertilizers.

Despite the major demand for biofuels for meeting future energy requirements, there are major hurdles during the implementation of the technology related to the biological conversion of biowaste. The efficiency of the biological processes involved in anaerobic digestion, microbial fermentation, dark fermentation, or transesterification, and the final yield of the corresponding biofuel depends on the metabolism and dynamics of the microbial community. There is a strict requirement for process control and optimization of process parameters like pH, temperature, nutrient availability, etc. for the growth and metabolism of the microbes. Another factor requiring special attention with respect to enhanced efficiency of the biotechnological processes is the pretreatment of the biowaste substrate for enabling its efficient hydrolysis which in turn is required for the action of the microbes to bring about biological degradation. In the case of biodiesel generation from biowastes, extraction of lipid or oil entails costly procedures. Also, the quality of oil extracted for e.g., the presence of free fatty acid and moisture content can affect the transesterification process adversely and hence the quality of biodiesel thus generated.

3.8 Conclusion

Biowastes are potential low-cost renewable sources for sustainable biotechnological transformation to bioenergy. Both laboratory scale and industrial level research have established the technological soundness for using biowastes for the generation of biofuels. Anaerobic digestion and microbial fermentation are mature biotechnological processes for yielding biofuels in the form of biogas, bioethanol, and biodiesel. Europe is the leading producer of biogas for use in generating electricity, heat, and transportation, thanks to the renewable energy policies adopted in addition to the environmental, economic, and climate benefits. Likewise, bioethanol and biodiesel production are recorded as the highest in the USA followed by Brazil, European Union, etc. Both bioethanol and biodiesel are used as transportation fuels replacing gasoline either completely or partially. Biodiesel production worldwide has seen an upward trend from 3.9 billion liters in 2005 to 18.1 billion liters in 2010 and is expected to reach 41.4 billion liters in 2025. The rapid expansion rate in the generation and usage of biofuels is mainly on account of the adoption of renewable energy policies by governments as well as incentives given to people. Support from government and world bodies have helped in lowering production costs and in enabling cost competitiveness of the biofuel-derived bioenergy [106]. More effort is required by developing countries for promoting the production and commercialization of biofuels from biowaste.

Acknowledgments

This work is supported by the Graphic Era University Dehradun

References

[1] Rashidi S, Esfahani JA, Hormozi F. (2020). Classifications of Porous Materials for Energy Applications. Reference Module in Materials Science and Materials Engineering. Encyclopedia of Smart Materials: Elsevier Inc. DOI: https://doi.org/10.1016/b978-0-12-803581-8.11739-4

[2] Directive 2008/98/EC of the European Parliament and of the Council of 19 November 2008 on Waste and Repealing Certain Directives. Available online: https://eur-lex.europa.eu/legal-content/EN/TXT/PDF/?uri=%20CELEX:32008L0098&from=EN%20(accessed%20on%206%20February%202019)

[3] https://www.epa.gov/facts-and-figures-about-materials-waste-and-recycling/national-overview-facts-and-figures-materials

[4] Gustafsson U, Wills W, Draper A. Food and public health: contemporary issues and future directions. Critic. Public Health. 2011; 21(4): 385–393. DOI: https://doi.org/10.1080/09581596.2011.625759

[5] Idehai IM, Akujieze CN. Estimation of landfill gas and its renewable energy potential in Lagos, Nigeria. Int J. Energy Environ. Eng. 2015; 6:329–343. DOI: https://doi.org/10.1007/s40095-015-0178-9

[6] McKendry P. Energy production from biomass (Part 2): conversion technologies. Bioresour. Technol. 2002; 83: 47–54. DOI: https://doi.org/10.1016/S0960-8524(01)00119-5

[7] Pham TPT, Kaushik R, Parshetti GK, Mahmood R, Balasubrmanian R. Food waste-to-energy conversion technologies: current status and future directions. Waste Manag. 2015; 38: 399–408. DOI: https://doi.org/10.1016/j.wasman.2014.12.004

[8] Karmee SK. Liquid biofuels from food waste: current trends, prospect, and limitation. Renew. Sustain. Energy Rev. 2016; 53: 945–953. DOI: https://doi.org/10.1016/j.rser.2015.09.041

[9] Murugan K, Chandrasekaran VS, Karthikeyan P, Al-Sohaibani S. 2013.The current state of the art of food processing by-products. In: Chandrasekaran, M. (Ed.), Valorisation of Food Processing By-products. Taylor and Francis Group, Florida, pp. 35–62. https://doi.org/10.1201/b12816

[10] Nayak A, Bhushan B. An overview of the recent trends in the waste valorization techniques for food wastes. J. Environ. Manag. 2019; 233: 352–370. DOI: https://doi.org/10.1016/j.jenvman.2018.12.041

[11] Murphy JD, McKeogh E, Kiely G. Technical/economic/environmental analysis of biogas utilization. Appl. Energy 2004; 77 (4): 407–427. DOI: https://doi.org/10.1016/j.apenergy.2003.07.005

[12] Pham TPT, Kaushik R, Parshetti GK, Mahmood R, Balasubrmanian R. Food waste-to-energy conversion technologies: current status and future directions. Waste Manag. 2015; 38: 399–408. DOI: https://doi.org/10.1016/j.wasman.2014.12.004

[13] Karmee SK. Liquid biofuels from food waste: current trends, prospect and limitation. Renew. Sustain. Energy Rev. 2016; 53: 945–953. DOI: https://doi.org/ 10.1016/j.rser.2015.09.041

[14] Zhang Q, Hu J, Lee D-J. Biogas from anaerobic digestion processes: research updates. Renew. Energy 2016a; 98: 108–119. DOI: https://doi.org/10.1016/j.renene.2016.02.029

[15] Fisgativa H, Tremier A, Dabert P. Characterizing the variability of food waste quality: a need for efficient valorization through anaerobic digestion. Waste Manag. 2016; 50: 264–274. DOI: https://doi.org/10.1016/j.wasman.2016.01.041

[16] Agdag ON, Sponza DT. Co-digestion of mixed industrial sludge with municipal solid wastes in anaerobic simulated landfilling bioreactors, J. Hazard. Mater. 2007; 140: 75-85. DOI: https://doi.org/10.1016/j.jhazmat.2006.06.059

[17] Gómez X, Cuetos MJ, Cara J, Morán A, García AI. Anaerobic co-digestion of primary sludge and the fruit and vegetable fraction of the municipal solid wastes. Renew. Energy 2006; 31 (12): 2017–2024. DOI: https://doi.org/10.1016/j.renene.2005.09.029

[18] Yan BH, Selvam A, Wong JWC. Application of rumen microbes to enhance food waste hydrolysis in acidogenic leach-bed reactors. Bioresour. Technol. 2014;168: 64–71. DOI: https://doi.org/10.1016/j.biortech.2014.03.085

[19] Zamanzadeh M, Hagen LH, Svensson K, Linjordet R, Horn SJ. Biogas production from food waste via co-digestion and digestion- effects on performance and microbial ecology. Sci. Rep. 2017; 7: 17664. DOI: https://doi.org/10.1038/s41598-017-15784-w

[20] Cesaro A, Belgiorno V. Pre-treatment methods to improve anaerobic biodegradability of organic municipal solid waste fractions. Chem Eng. J. 2014; 240: 24–37. DOI: https://doi.org/10.1016/j.cej.2013.11.055

[21] Ariunbaatar J, Panico A, Esposito G, Pirozzi F, Lens PNL. Pre-treatment methods to enhance anaerobic digestion of organic solid waste, Appl. Energy. 2014; 123: 143–156. DOI: https://doi.org/10.1016/j.apenergy.2014.02.035

[22] Carlsson M, Lagerkvist A, Morgan-Sagastume F. The effects of substrate pre-treatment on anaerobic digestion systems: a review. Waste Manage. 2012; 32: 1634–1650. DOI: https://doi.org/10.1016/j.wasman.2012.04.016

[23] Cesaro A, Belgiorno V. Sonolysis and ozonation as pre-treatment for anaerobic digestion of solid organic waste. Ultrason. Sonochem. 2013; 20: 931–936. DOI: https://doi.org/10.1016/j.ultsonch.2012.10.017

[24] Ferrer I, Ponsá S, Vázquez F, Font X. Increasing biogas production by thermal (70 °C) sludge pre-treatment prior to thermophilic anaerobic digestion. Biochem. Eng. J. 2008; 42: 186–192. DOI: https://doi.org/10.1016/j.bej.2008.06.020

[25] Izumi K, Okishio Y, Nagao N, Niwa C, Yamamoto S, Toda T. Effects of particle size on anaerobic digestion of food waste. Int. Biodeterior. Biodegr. 2010; 64: 601–608. DOI: https://doi.org/10.1016/j.ibiod.2010.06.013

[26] Jackowiak D, Sfrigon JC, Ribeiro T, Pauss A, Guiot S. Enhancing solubilization and methane production kinetic of switchgrass by microwave pre-treatment. Bioresour. Technol. 2011;102: 3535–3540. DOI: https://doi.org/10.1016/j.biortech.2010.11.069

[27] Tiehm A, Nickel K, Zellhorn M, Neis U. Ultrasonic waste activated sludge disintegration for improving anaerobic stabilization. Water Res. 2001; 35: 2003–2009. DOI: https://doi.org/10.1016/s0043-1354(00)00468-1

[28] Fdéz-Güelfo A, Álvarez-Gallego C, Sales-Márquez D, Romero García LI. Biological pre-treatment applied to industrial organic fraction of municipal solid waste (OFMSW): Effect on anaerobic digestion. Chem. Eng. J. 2011; 172: 321–325. DOI: https://doi.org/10.1016/j.cej.2011.06.010

[29] Lina L, Xu F, Ge X, Li Y. Improving the sustainability of organic waste management practices in the food-energy-water nexus: A comparative review of anaerobic digestion and composting. Renew. Sustain. Energy Rev. 2018; 89: 151–167. DOI: https://doi.org/10.1016/j.rser.2018.03.025

[30] Salama El-S, Saha SS, Kurade MB, Dev S, Chang SW, Jeon B-H. Recent trends in anaerobic co-digestion: Fat, oil, and grease (FOG) for enhanced biomethanation. Prog. Energy Combust. Sci. 2019; 70: 22–42. DOI: https://doi.org/10.1016/j.pecs.2018.08.002

[31] Mao C, Feng Y, Wang X, Ren G. Review on research achievements of biogas from anaerobic digestion. Renew. Sustain. Energy Review. 2015; 45: 540–555. DOI: https://doi.org/10.1016/j.rser.2015.02.032

[32] Singh M, Srivastava R. Sequencing batch reactor technology for biological wastewater treatment: a review. Asia-Pac J. Chem. Eng. 2011; 6: 3–13. DOI: https://doi.org/10.1002/apj.490

[33] Boe K, Angelidaki I. Serial CSTR digester configuration for improving biogas production from manure. Water Res. 2009; 43:166–72. DOI: https://doi.org/10.1016/j.watres.2008.09.041

[34] Singh S, Prerna P. Review of recent advances in anaerobic packed-bed biogas reactors. Renew. Sustain. Energy Rev. 2009; 13: 1569–1575. DOI: https://doi.org/10.1016/j.rser.2008.08.006

[35] Şentürk E, Ýnce M, Engin GO. The effect of transient loading on the performance of a mesophilic anaerobic contact reactor at constant feed strength. J. Biotechnol. 2012; 164:232–237. DOI: https://doi.org/10.1016/j.jbiotec.2012.08.016

[36] Ware A, Power N. Biogas from cattle slaughterhouse waste: Energy recovery towards an energy self-sufficient industry in Ireland. Renewable Energy, (2016; 97: 541- 549. DOI: https://doi.org/10.1016/j.renene.2016.05.068

[37] Parawira W, Murto M, Zvauya R, Mattiasson B. Anaerobic batch digestion of solid potato waste alone and in combination with sugar beet leaves. Renew. Energy 2004; 29 (11): 1811–1823. DOI: https://doi.org/10.1016/j.renene.2004.02.005

[38] Kim JK, Oh BR, Chun YN, Kim SW. Effects of temperature and hydraulic retention time on anaerobic digestion of food waste. J. Biosci. Bioeng. (2006;102(4): 328–332. DOI: https://doi.org/10.1263/jbb.102.328

[39] Parawira W, Murto M, Read JS, Mattiasson B. Profile of hydrolases and biogas production during two-stages mesophilic anaerobic digestion of

solid potato waste. Process Biochem. 2005; 40 (9): 2945–2952. DOI: https://doi.org/10.1016/j.procbio.2005.01.010

[40] Kim JK, Han GH, Oh BR, Chun YN, Eom, C-Y, Kim SW. Volumetric scale-up of a three-stage fermentation system for food waste treatment. Bioresour. Technol. (2008; 99(10): 4394 -4399. DOI: https://doi.org/10.1016/j.biortech.2007.08.031

[41] Zhang C, Xiao G, Peng L, Su H, Tan T. The anaerobic co-digestion of food waste and cattle manure. Bioresour. Technol. 2013; 129: 170–176. DOI: https://doi.org/10.1016/j.biortech.2012.10.138

[42] Alvarez R, Lidén G. Semi-continuous co-digestion of solid slaughterhouse waste, manure, and fruit and vegetable waste. Renew. Energy. 2008; 33(4): 726–734. DOI: https://doi.org/10.1016/j.renene.2007.05.001

[43] Banks CJ, Chesshire M, Heaven S, Arnold R. Anaerobic digestion of source-segregated domestic food waste: Performance assessment by mass and energy balance. Bioresour. Technol. (2011; 102(2): 612–620. DOI: https://doi.org/10.1016/j.biortech.2010.08.005

[44] Neves L, Oliveira R, Alves MM. Anaerobic co-digestion of coffee waste and sewage sludge. Waste Manag. 2006; 26(2): 176–181. DOI: https://doi.org/10.1016/j.wasman.2004.12.022

[45] Li Y, Cheng H, Guo G, Zhang T, Qin Y, Li YY. High solid mono-digestion and co-digestion performance of food waste and sewage sludge by a thermophilic anaerobic membrane bioreactor. Bioresour. Technol. 2020; 310:123433. DOI: https://doi.org/10.1016/j.biortech.2020.123433

[46] Gaoab M, Yanga M, Ma X, Xie D, Wu C, Wang Q. Effect of co-digestion of tylosin fermentation dreg and food waste on anaerobic digestion performance. Bioresour. Technol. 2021; 325: 124693. DOI: https://doi.org/10.1016/j.biortech.2021.124693

[47] Toyohashi City Waterworks & Sewarage Bureau, B. recycle in T.C., 2019. Biomass recycle in Toyohashi City. Toyohashi.

[48] Kim S, Dale BE. Global potential bioethanol production from wasted crops and crop residues. Biomass Bioenergy. 2004; 26: 361–375. DOI: https://doi.org/10.1016/j.biombioe.2003.08.002

[49] Kefale A, Redi M, Asfaw A. Potential of bioethanol production and optimization test from agricultural waste: the case of wet coffee processing waste (pulp). Int. J. Renew. Energy Resour. 2012; 2: 446–450. DOI: https://www.researchgate.net/publication/260310759

[50] Sharma N, Kalra KL, Oberoi HS, Bansal S. Optimization of fermentation parameters for production of ethanol from kinnow waste and banana peels by simultaneous saccharification and fermentation. Indian J. Microbiol. 2007; 47 (4): 310–316. DOI: https://doi.org/10.1007/s12088-007-0057-z

[51] Kim JK, Oh BR, Shin HJ, Eomd CY, Kim SW. Statistical optimization of enzymatic saccharification and ethanol fermentation using food waste. Process Biochem. 2008b; 43 (11): 1308–1312. DOI: https://doi.org/10.1016/j.procbio.2008.07.007

[52] Hong YS, Yoon HH. Ethanol production from food residues. Biomass and Bioenergy. 2011; 35(7): 3271-3275. DOI: https://doi.org/10.1016/j.biombioe.2011.04.030

[53] Saha BC, Cotta MA. Lime pretreatment, enzymatic saccharification and fermentation of rice hulls to ethanol. Biomass and Bioenergy. 2008; 32(10): 971–977. DOI: https://doi.org/10.1016/j.biombioe.2008.01.014

[54] Kim JH, Lee JC, Pak D. Feasibility of producing ethanol from food waste. Waste Manag. 2011; 31(9-10): 2121-2125. DOI: https://doi.org/10.1016/j.wasman.2011.04.011

[55] Oberoi HS, Vadlani PV, Saida L, Bansal S, Hughes JD. Ethanol production from banana peels using statistically optimized simultaneous saccharification and fermentation process. Waste Manag. 2011; 31(7): 1576–1584. DOI: https://doi.org/10.1016/j.wasman.2011.02.007

[56] Oberoi, HS, Vadlani PV, Nanjundaswamy A, Bansal S, Singh S, Kaur S, Babbar N. Enhanced ethanol production from Kinnow mandarin (Citrus reticulata) waste via a statistically optimized simultaneous saccharification and fermentation process. Bioresour. Technol. (2011;102(2): 1593–1601. DOI: https://doi.org/10.1016/j.biortech.2010.08.111

[57] Magyar M, da Costa Sousa L, Jayanthi S, Balan V. Pie waste – A component of food waste and a renewable substrate for producing ethanol. Waste Manag. 2017; 62: 247–254. DOI: https://doi.org/10.1016/j.wasman.2017.02.013

[58] Bibra M, Rathinam NK, Johnson GR, Sani RK. Single pot biovalorization of food waste to ethanol by Geobacillus and Thermoanaerobacter spp. Renewable Energy. 2020; 155:1032-1041. DOI: https://doi.org/10.1016/j.renene.2020.02.093

[59] Belkacemi K, Turcotte G, Savoie P. Aqueous/Steam-Fractionated Agricultural Residues as Substrates for Ethanol Production. Ind. Eng. Chem. Res. 2002;41(2): 173–179. DOI: https://doi.org/10.1021/ie0102246

[60] Yang X, Lee SJ, Yoo HY, Choi HS, Park C, Kim SW. Biorefinery of instant noodle waste to biofuels. Bioresour. Technol. 2014; 159:17–23. DOI: https://doi.org/10.1016/j.biortech.2014.02.068

[61] Chandel AK, Kapoor RK, Singh A, Kuhad RC. Detoxification of sugarcane bagasse hydrolysate improves ethanol production by Candida shehatae NCIM 3501. Bioresour. Technol. 2007; 98: 1947–1950. DOI: https://doi.org/10.1016/j.biortech.2006.07.047

[62] Sukumaran RK, Singhania RR, Mathew GM, Pandey A. Cellulase production using biomass feed stock and its application in lignocellulose saccharification for bio-ethanol production. Renew. Energy. 2009; 34(2): 421–424. DOI: https://doi.org/10.1016/j.renene.2008.05.008

[63] Liu Z-H, Qin L, Zhu J-Q, Li B-Z, Yuan Y-J. Simultaneous saccharification and fermentation of steam-exploded corn stover at high glucan loading and high temperature. Biotechnology for Biofuels, (2014; 7(1): 167. DOI: https://doi.org/10.1186/s13068-014-0167-x

[64] Alfani F, Gallifuoco A, Saporosi A, Spera A, Cantarella M. Comparison of SHF and SSF processes for the bioconversion of steam-exploded wheat straw. J. Indus. Microbiol. Biotechnol.2000; 25(4): 184–192. DOI: https://doi.org/10.1038/sj.jim.7000054

[65] Ban-Koffi L, Han YW. Alcohol production from pineapple waste. World J. Microbiol. Biotechnol. 1990; 6: 281–284. DOI: https://doi.org/10.1007/BF01201297

[66] Korkie LJ, Janse BJH, Viljoen-Bloom M. Utilizing grape pomace for ethanol production. S. Afr. J. Enol. Vitic. 2002; 23: 31–36. DOI: https://doi.org/10.21548/23-1-2152

[67] Chong M-L, Sabaratnam V, Shirai Y, Hassan MA. Biohydrogen production from biomass and industrial wastes by dark fermentation. Int. J. Hydrog. Energy. 2009; 34(8): 3277–3287. DOI: https://doi.org/10.1016/j.ijhydene.2009.02.010

[68] Priya S, Brijesh B, Reddy KR, Reddy CV, Shetti NP, Kulkarni RV, Raghu AV. Prospects of Biohydrogen Production from Organic Waste - A Review. Chem. Eng. Technol. 2020; 43(7): 1240–1248. DOI: https://doi.org/10.1002/ceat.201900400

[69] Kamaraj M, Ramachandran KK, Aravind J. Biohydrogen production from waste materials: benefits and challenges. Int. J. Environ. Sci. Technol. 2019. DOI: https://doi.org/10.1007/s13762-019-02577-z

[70] Van Ginkel SW, Oh SE, Logan BE. Biohydrogen gas production from food processing and domestic wastewaters. Int. J. Hydro. Energy. 2005; 30(15): 1535–1542. DOI: https://doi.org/10.1016/j.ijhydene.2004.09.017

[71] Han, S. Biohydrogen production by anaerobic fermentation of food waste. Int. J. Hydrog. Energy. 2004; 29(6): 569–577. DOI: https://doi.org/10.1016/j.ijhydene.2003.09.001

[72] Pan J, Zhang R, Elmashad H, Sun H, Ying Y. Effect of food to microorganism ratio on biohydrogen production from food waste via anaerobic fermentation. Int. J. Hydrog. Energy. 2008; 33(23): 6968–6975. DOI: https://doi.org/10.1016/j.ijhydene.2008.07.130

[73] Fand H, Li C, Zhang T. Acidophilic biohydrogen production from rice slurry. Int. J. Hydrog. Energy. 2006; 31(6): 683–692. DOI: https://doi.org/10.1016/j.ijhydene.2005.07.005

[74] Zhang T, Liu H, Fang HHP. Biohydrogen production from starch in wastewater under thermophilic conditions. J. Environ. Manag. 2003; 69(2) 149–156. DOI: https://doi.org/10.1016/s0301-4797(03)00141-5

[75] O-Thong S, Prasertsan P, Intrasungkha N, Dhamwichukorn S, Birkeland N-K. Improvement of biohydrogen production and treatment efficiency on palm oil mill effluent with nutrient supplementation at thermophilic conditions using an anaerobic sequencing batch reactor. Enzyme Microb. Technol. (2007; 41(5): 583–590. DOI: https://doi.org/10.1016/j.enzmictec.2007.05.002

[76] Ferchichi M, Crabbe E, Gil G-H, Hintz W, Almadidy A. Influence of initial pH on hydrogen production from cheese whey. J. Biotechnol. 2005; 120(4): 402–409. DOI: https://doi.org/10.1016/j.jbiotec.2005.05.017

[77] Venkata Mohan S, Lalit Babu V, Sarma PN. Anaerobic biohydrogen production from dairy wastewater treatment in sequencing batch reactor (AnSBR): effect of organic loading rate. Enzyme Microb. Technol. 2007; 41:506–515. DOI: https://doi.org/10.1016/j.enzmictec.2007.04.007

[78] Kim DH, Kim SH, Kim KY, Shin HS. Experience of a pilot-scale hydrogen-producing anaerobic sequencing batch reactor (ASBR) treating food waste. Int. J. Hydrog. Energy. 2010; 35(4): 1590-1594. DOI: https://doi.org/10.1016/j.ijhydene.2009.12.041

[79] Yun Y-M, Lee M-K, Im S-W, Marone A, Trably E, Shin S-R, Kim D-H. Biohydrogen production from food waste: Current status, limitations, and future perspectives. Bioresour. Technol. 2018; 248: 79–87. DOI: https://doi.org/10.1016/j.biortech.2017.06.107

[80] Jarunglumlert T, Prommuak C, Putmai N, Pavasant P. Scaling-up bio-hydrogen production from food waste: Feasibilities and challenges. Int. J. Hydrog. Energy. 2018; 43: 634–648. DOI: https://doi.org/10.1016/j.ijhydene.2017.10.013

[81] Kiran EU, Trzcinski AP, Ng WJ, Liu Y. Bioconversion of food waste to energy: a review. Fuel 2014;134: 389–399. DOI: https://doi.org/10.1016/j.fuel.2014.05.074

[82] Cavinato C, Giuliano A, Bolzonella D, Pavan P, Cecchi F. Biohythane production from food waste by dark fermentation coupled with anaerobic digestion process: a long-term pilot scale experience. Int. J. Hydrog. Energy. 2012;37 (15): 11549–11555. DOI: https://doi.org/10.1016/j.ijhydene.2012.03.065

[83] Shimada Y, Watanabe Y, Sugihara A, Tominaga Y. "Enzymatic Alcoholysis for Biodiesel Fuel Production and Application of the Reaction to Oil Processing," Journal of Molecular Catalysis B: Enzymatic, 2002; 17(3-5): 133-142. DOI: https://doi.org/10.1016/S1381-1177(02)00020-6

[84] Jain SK, Kumar S, Chaube A. Technical sustainability of biodiesel and its blends with diesel in CI engines: a Review. Int. J. Chem. Eng. Appl. 2011; 2:101. DOI: http://ijcea. org/papers/84-A565.pdf.

[85] Ma FR, Hanna MA. Biodiesel Production: A Review, Bioresour. Technol. 1999; 70(1):1-15. DOI: https://doi.org/10.1016/S0960-8524(99)00025-5

[86] Karmee SK. Liquid biofuels from food waste: Current trends, prospect and limitation. Renew. Sustain. Energy Review. 2016; 53: 945–953. DOI: http://doi:10.1016/j.rser.2015.09.041

[87] Muniyappa PR, Brammer SC, Noureddini N. Improved Conversion of Plant Oils and Animal Fats into Biodiesel and Co-Product, Bioresour. Technol. 1996;56(1):19-24. DOI: https://doi.org/10.1016/0960-8524(95)00178-6

[88] Leadbeater NE, Stencel LM. Fast, Easy Preparation of Biodiesel Using Microwave Heating. Energy Fuels. 2006; 20 (5): 2281-2283. DOI: https://doi.org/10.1021/ef060163u

[89] Refaat AA, Sheltawy ST, Sadek KU. Optimum Reaction Time, Performance and Exhaust Emissions of Biodiesel Produced by Microwave Irradiation. Int. J. Environ. Sci. Technol. 2008; 5 (3):315-322. DOI: https://doi.org/10.1007/BF03326026

[90] Patil P, Veera G, Mannarswamy A, Deng S, Cooke P, Munson-McGee S, Lammers P, Khandan N. Optimization of Microwave-Assisted Transesterification of Dry Algal Biomass Using the Response Surface Methodology. Bioresour. Technol. 2011; 102(2):1399-1405. DOI: https://doi.org/10.1016/j.biortech.2010.09.046

[91] Azcan N, Danisman A. Alkali Catalyzed Transesterification of Cottonseed Oil by Microwave Irradiation. Fuel. 2007; 86(17-18): 2639-2644. DOI: https://doi.org/10.1016/j.fuel.2007.05.021

[92] Kayode B, Hart A. An overview of transesterification methods for producing biodiesel from waste vegetable oils. Biofuels. 2017:1–19.DOI: https://doi.org/10.1080/17597269.2017.1306683

[93] Patil PD, Gude VG, Reddy HK, Muppaneni T, Deng S. Biodiesel production from waste cooking oil using sulfuric acid and microwave irradiation processes. J. Environ. Protect. 2012; 3: 107. DOI: https://doi.org/10.4236/jep.2012.31013.

[94] Kawentar WA, Budiman A. Synthesis of Biodiesel from Second-Used Cooking Oil. Energy Procedia, 2013; 32: 190–199. DOI: https://doi:10.1016/j.egypro.2013.05.025

[95] Kaewdaeng S, Sintuya P, Nirunsin R. Biodiesel production using calcium oxide from river snail shell ash as a catalyst. Energy Procedia, 2017;138: 937–942. DOI: https://doi:10.1016/j.egypro.2017.10.057

[96] Chung K-H, Kim J, Lee K-Y. Biodiesel production by transesterification of duck tallow with methanol on alkali catalysts. Biomass Bioenergy. 2009; 33: 155–8. DOI: https://doi.org/10.1016/j.biombioe.2008.04.014

[97] Mata TM, Cardoso N, Ornelas M, Neves S, Caetano NS. Evaluation of two purification methods of biodiesel from beef tallow, pork lard, and chicken fat. Energy Fuels 2011; 25: 4756–62. https://doi.org/10.1021/ef2010207

[98] Cunha A, Feddern V, De Pr MC, Higarashi MM, de Abreu PG, Coldebella A. Synthesis and characterization of ethylic biodiesel from animal fat wastes. Fuel. 2013; 105: 228–34. DOI: https://doi.org/10.1016/j.fuel.2012.06.020

[99] Madhu D, Singh B, Sharma YC. Studies on the application of fish waste for the synthesis of high-quality biodiesel. RSC Adv 2014; 4: 31462–8. DOI: https://doi.org/10.1039/C4RA03590A

[100] Arumugam A, Ponnusami V. Production of biodiesel by enzymatic transesterification of waste sardine oil and evaluation of its engine performance. Heliyon 2017; 3: e00486. DOI: https://doi.org/10.1016/j.heliyon.2017.e00486

4

Coconut Shell-Based Activated Carbon Supported Metal Oxides in Catalytic Cracking Activity

Tavayogeshwary Thangadurai and Ching Thian Tye

School of Chemical Engineering, Universiti Sains Malaysia, 14300 Nibong Tebal, Pulau Pinang, Malaysia
E-mail: chcttye@usm.my

Abstract

Activated carbon-based metal oxides were tested for their performance in catalytic cracking of waste cooking oil to investigate the influence of their properties on the catalyst activity. The activated carbon produced a higher liquid yield with more C_5–C_{20} hydrocarbon yields compared to commonly used ZSM-5 and Al_2O_3, in a continuous fixed bed reactor system. Coconut shell-based activated carbon showed better catalytic performance than wood-based activated carbon due to its higher surface area and more porous structure. In addition, metal oxides were incorporated onto activated carbon via the incipient wetness impregnation method to provide additional active sites for the reaction. Despite the increased liquid yield, the C_5–C_{20} hydrocarbon yield varied in the sequence of metal: Mo $<$ Cu $<$ Ni $<$ Co $<$ Fe oxides supported on activated carbon. Change in strength and density of the catalyst acid and basic sites produced high C_{15} and C_{17} hydrocarbons contributed by better deoxygenation via decarboxylation and decarbonylation with catalytic cracking of waste cooking oil.

Keywords: Renewable energy, catalytic cracking, waste cooking oil, green fuel, activated carbon, and metal oxide.

4.1 Introduction

Green fuel produced via catalytic cracking of vegetable oil has lower oxygen content, which resulted in better thermal and oxidation stability, compared to biodiesel (fatty acid methyl ester) produced from transesterification. During a catalytic cracking process, triglycerides are decomposed to various oxygen compounds such as carboxylic acids, ester, aldehyde, etc. These oxygen components and fatty acids are then converted to hydrocarbon through deoxygenation via decarbonylation or decarboxylation [1]. Catalysts serve to maximize the conversion of triglycerides and boost desired hydrocarbon yield of the reaction. Heterogeneous metal-based catalysts such as metal sulfides, metal phosphides, metal carbides, metal oxides, and reduced metal are the solid acid catalysts commonly used for deoxygenation. Sulfided Ni-W, Ni-Mo, and Co-Mo catalysts exhibited high diesel range paraffin yields via decarboxylation, and decarbonylation were reported in the cracking of vegetable oil in the presence of hydrogen [2]. Despite its high selectivity for hydrocarbon production, the reaction product was contaminated by sulfur leaching of the sulfided catalysts. On the other hand, phosphide and carbide catalysts have complex synthesis processes and short lifetimes due to the transformation of their active phases during a reaction. Although high cracking activity was also reported with reduced metal catalysts at mild temperature with minimal requirement of hydrogen supply [3], water produced via decarbonylation and hydrodeoxygenation during reaction reduced the stability and activity of the catalysts [4]. Metal oxide catalysts are relatively favored in catalytic cracking reactions as they boost deoxygenation and reduce product decomposition into gaseous yield [5].

The surface area and porous structure of a catalyst play important role in reaction activity and catalyst stability. Activated carbon generally possessed high porosity (0.5–1.4 cm^3/g) and surface area (600–1400 m^2/g) [9]. Higher C_{15} and C_{17} hydrocarbons yield was reported with activated carbon as catalyst support compared to those obtained by using common catalyst support: ZrO_2, SiO_2, or Al_2O_3 [7]. High catalyst stability with resistance to coke formation was also reported exhibited by waste-derived carbon-based catalysts in the production of diesel-range hydrocarbons in deoxygenation reaction [8]. Activated carbon can be prepared by carbonizing coconut shell, wood, coal, peat, and/or lignite under nitrogen flow at a temperature higher than 700 °C. Activation can be done physically using an oxidizing gas such as steam, CO_2, and air or chemically with acid, alkali, and a metal salt. Characteristics of activated carbon also can be tuned towards polar or nonpolar substances.

For instance, active sites with a stronger affinity toward polar substances are provided by oxygen functional groups, and highly porous structures are developed using acid treatment [9]. The porous structure is manipulated by the nature of the source material, preparation method, preparation conditions, and modifications [10]. Metal oxides on activated carbon-supported catalysts act as an active phase for their acid-base and redox properties [11]. 65% hydrocarbon yield with 12% n-C_{15} selectivity was obtained with carbon-supported MgO. Hydrocarbon yield of 72–96% with mainly n-(C_{15} + C_{17}) selectivity in the range of 82–93% were reported in catalytic cracking of vegetable oil with activated carbon-supported CaO, NiO, Co_3O_4, Ag_2O_3, and La_2O_3 [1].

Porous structure and acidity or basicity of active sites of a catalyst influence its deoxygenation performance in catalytic cracking reaction [12]. The present study aims to investigate the catalytic cracking of waste cooking oil using different activated carbon-supported transition metal oxide as a catalyst in order to gain a better understanding of the relations between the catalyst properties and the product yields and selectivities.

4.2 Experimental Procedures

4.2.1 Material

Waste cooking oil was collected from the cafeteria in Universiti Sains Malaysia, Nibong Tebal, Malaysia. The waste cooking oil was filtered for impurities removal and heated to eliminate moisture. Commercially available wood-based and coconut shell-based activated carbon powder was obtained from KI Carbon Solutions. Nitric acid (69–70%) was obtained from QREC. Nickel (II) nitrate hexahydrate ($Ni(NO_3)_2 \bullet 6H_2O$) with purity 99.999%, Cobalt (II) nitrate hexahydrate ($Co(NO_3)_2 \bullet 6H_2O$) with purity 99.999%, Ammonium heptamolyb-date tetrahydrate ($(NH_4)_6Mo_7O_{24} \bullet 4H2O$) with purity 99.98%, Copper (II) nitrate hydrate ($Cu(NO_3)_2 \bullet xH_2O$) with purity 99.999% and Iron (III) nitrate nonahydrate ($Fe(NO_3)_3 \bullet 9H_2O$) with purity 99.999% were obtained from Sigma-Aldrich.

4.2.2 Catalyst Preparation

Two types of activated carbon: wood-based (w-AC) and coconut shell-based (c-AC) were used. They were pre-treated with boiling nitric acid (69–70% QREC) solution [13]. The activated carbons were then impregnated with a 10 wt. % of specific metal using the incipient wetness impregnation method.

The transition metals investigated were Ni, Co, Mo, Fe, and Cu. The metal precursors were dissolved in deionized water and the aqueous solution was stirred with activated carbon overnight. The catalysts prepared were dried at 105 °C in an oven overnight and calcined at 550 °C for 5 hours in nitrogen flow (99.9995%) at 30 ml/min before reaction.

4.2.3 Catalytic Cracking of Waste Cooking Oil

Catalytic cracking of waste cooking oil was carried out in a tubular fixed-bed reactor (Figure 4.1). 1 g catalyst was loaded on top of the flattened quartz wool inside the reactor. Nitrogen (99.9995%) as carrier gas was passed through the reactor via a mass flow controller (model M100B by MKS Instrument) set at 30 ml/min. The temperature of the reactor was gradually raised at a heating rate of 10 °C/minute to 450 °C in a tube furnace (Nabertherm

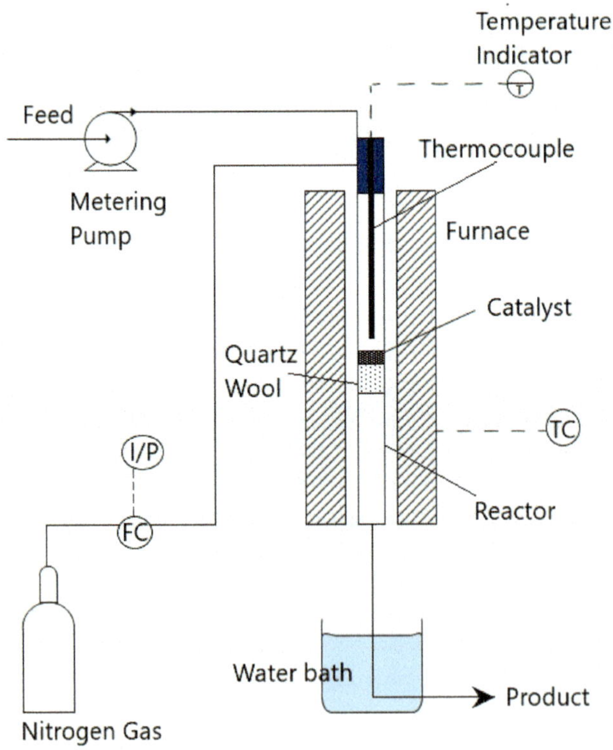

Figure 4.1 Experimental setup for catalytic cracking of waste cooking oil in a fixed bed reactor

B170). Once the temperature achieved 450 °C, waste cooking oil was charged as feed (WHSV 9 hour^{-1}) into the reactor using a Series I high-performance metering pump by Lab Alliance and the reaction time was considered started. The reaction was carried out for 60 minutes and the liquid product was collected at the outlet during this period. The reactor was left to cool down in nitrogen gas flow to room temperature after the reaction. The used catalyst was then recovered for analysis.

4.2.4 Product Analysis

The waste cooking oil feed was analyzed using Agilent 7890A gas chromatography equipped with mass spectroscopy (GC-MS) and an HP-5MS column (30 m x 0.25 mm x 0.25μm). Liquid products were analysed using a GC equipped with a flame ionization detector (GC-FID) and an HP-5 column (30 m $\times$ 0.32 mm $\times$ 0.25 μm). FID was operated at 300 °C with an injection temperature of 250 °C. The oven temperature was held at 40 °C for 6 minutes and then ramped to 270 °C at a heating rate of 7 °C/min which was followed by holding for 5 minutes. Helium (99.999%) at 1 ml/min serves as carrier gas and the liquid samples were diluted with hexane before being injected into GC. The liquid product was calculated for its yield and selectivity as follows.

$$Liquid\ yield\ (wt.\ \%) = \frac{Weight\ of\ liquid\ product\ (g)}{Weight\ of\ oil\ feed\ (g)} \times 100\% \tag{4.1}$$

$$Coke\ yield\ (wt.\%) = \frac{Used\ catalyst\ (g)\ -\ Fresh\ catalyst\ (g)}{Weight\ of\ oil\ feed\ (g)} \times 100\% \tag{4.2}$$

$$Gas\ yield\ (wt.\%) = 100\ -\ Liquid\ yield\ (wt.\%) - Coke\ yield\ (wt.\%) \tag{4.3}$$

$$Hydrocarbon\ yield(\%) = \frac{Area\ of\ C_5 - C_{20}\ hydrocarbons}{Total\ area - Area\ of\ hexane} \times 100\% \tag{4.4}$$

$$C_n\ selectivity(\%) = \frac{Area\ of\ C_n}{Area\ of\ C_5 - C_{20}\ hydrocarbons} \times 100\% \tag{4.5}$$

where ***n*** ranges from 5 to 20.

In addition, elemental analysis (carbon, hydrogen, nitrogen, and sulfur) with a relative standard deviation of 1 to 3% was carried out for the liquid product to determine its oxygen composition by difference [2].

4.3 Results and Discussion

4.3.1 Properties of Waste Cooking Oil

The fatty acid constituents of waste cooking oil were identified using gas chromatography-mass spectrometry. There were two major components observed which were oleic acid, $C_{18}H_{34}O_2$ (62.97%), and palmitic acid, $C_{16}H_{32}O_2$ (37.03%). These compounds could exist as free fatty acids or could be bonded by glycerol as a triglyceride molecule.

This result complies with the fatty acid composition of palm oil [15] as Malaysia's main cooking oil constituent. During frying, some triglyceride molecules are dissociated via hydrolysis caused by moisture in the food in the presence of heat. Therefore, there was a relatively higher free fatty acid content in used cooking oil compared to fresh vegetable oil with free fatty acids below 1%. Generally, waste cooking oil in Malaysia contains 18.4% of free fatty acids [1]. The high free fatty acid content in vegetable oil was anticipated to favor C–C cracking with lower deoxygenation via C–O scission [16] which produces lighter hydrocarbons. Oxygen groups in C_{16} and C_{18} fatty acids affect the oxidative stability and cold flow properties of the oil as fuel. Hence, waste cooking oil is required to be cracked into hydrocarbon chains in the fuel range and deoxygenated for better liquid product properties.

4.3.2 Catalytic Cracking of Waste Cooking Oil

4.3.2.1 Activated carbon-based catalysts

Catalyst activities of activated carbon derived from a different source (wood-based or coconut shell-based) and common catalyst bases: ZSM-5 and Al_2O_3 were compared. Catalytic cracking of waste cooking oil was conducted at reaction temperature 450 °C and oil feed rate of WHSV 9 hour^{-1} for 60 minutes. Liquid and coke yields were based on the amount collected during the first 60 mins of reaction while the gas yield was calculated through mass balance. The results in terms of the liquid, gas, and coke yields obtained are given in Table 4.1.

Most of the catalytic cracking processes gave a liquid yield higher than 44 wt. % except for catalytic cracking using ZSM-5 which gave 40.35 wt. % liquid yield. The coke formation (2.87 wt. %) and the gas yield (56.78 wt. %) were also the highest for the reaction with ZSM-5 in the current study. This could be due to the high acidity in ZSM-5 that promoted secondary reactions such as extensive cracking to form lighter compounds (gas products) and polymerization to generate residual substances that deposit on the catalyst

Table 4.1 Liquid, coke, and gas yields for catalytic cracking of waste cooking oil at 450 °C; WHSV 9 hr-1 and 60 mins.

The catalyst used in cracking reaction	Yield (wt. %)		
	Liquid	Coke	Gas
–	45.09	0.33	54.58
ZSM-5	40.35	2.87	56.78
Alumina	45.20	1.02	53.78
Wood Activated Carbon	44.48	1.16	54.36
Coconut Shell Activated Carbon	45.65	1.20	53.15

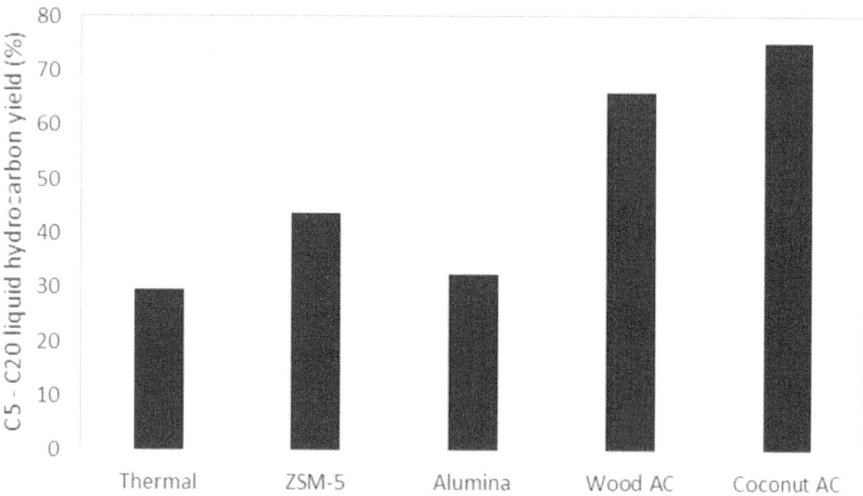

Figure 4.2 Liquid hydrocarbon yield obtained for thermal and catalytic cracking of waste cooking oil at 450 °C; WHSV 9 hr^{-1} and 60 mins.

as coke [17]. Al_2O_3 contributed to a less coke formation (1.02 wt. %) at a liquid yield of 45.20 wt. %. Meanwhile, wood-based and coconut shell-based activated carbons gave almost similar liquid yield (44–45 wt. %) with slightly higher coke deposition (~1.2 wt. %) than Al_2O_3 but lower than that of ZSM-5 was recorded. Liquid products were then further analyzed for their respective C_5–C_{20} liquid hydrocarbon yield as shown in Figure 4.2.

Despite the similar liquid yield range of around 44–46 wt. % obtained in thermal cracking and catalytic cracking over alumina, liquid hydrocarbon yield was found to be much lower at around 30% compared to that in liquid products (>60%) obtained in reactions with activated carbon as the catalyst. ZSM-5 generated a lower liquid yield, but the liquid product had a higher C_5–C_{20} liquid hydrocarbon yield (43%) than that produced using alumina

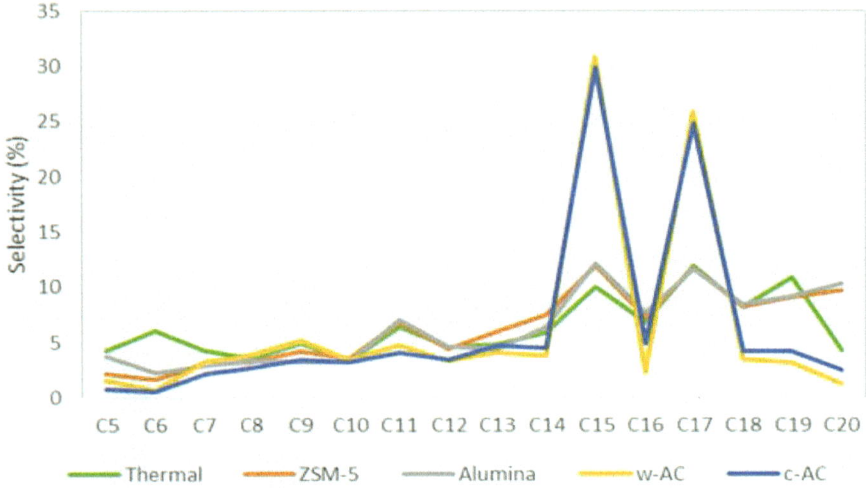

Figure 4.3 Carbon number selectivity for C_5–C_{20} hydrocarbons in thermal and catalytic cracking of waste cooking oil at 450 °C; WHSV 9 hr^{-1} and 30 mins.

(32%). Notably, activated carbons yielded more C_5–C_{20} liquid hydrocarbons than the other catalysts in the present study. Coconut shell-based and wood-based activated carbons gave a high liquid hydrocarbon yield of 75% and 66%, respectively.

Figure 4.3 shows carbon number selectivity for C_5–C_{20} hydrocarbons in liquid products. Liquid products exhibited relatively significant selectivity for n-(C_{15} + C_{17}) hydrocarbons. This could be deduced as the consequence of decarboxylation and/or decarbonylation of palmitic acid (C_{16}) and oleic acid (C_{18}) in WCO [1]. ZSM-5 and alumina were more selective toward producing heavier hydrocarbons. Both the wood-based and coconut shell-based activated carbon showed higher selectivity for C_{15} and C_{17} hydrocarbon than the other catalysts. This can be deduced that activated carbon has better deoxygenation and cracking activity towards lighter fuel range hydrocarbons.

The performance of a catalyst is directly related to its properties. Diffusion and transport of molecules in the heterogeneous catalytic reaction are influenced by the specific surface area and porosity of the catalyst. In the present study, the calcination of the activated carbon-based catalysts at high temperatures (550 °C) facilitates their higher surface area with the removal of impurities [27] for better porosity. BET isotherms for wood-based and coconut shell-based activated carbon are given in Figure 4.4(a).

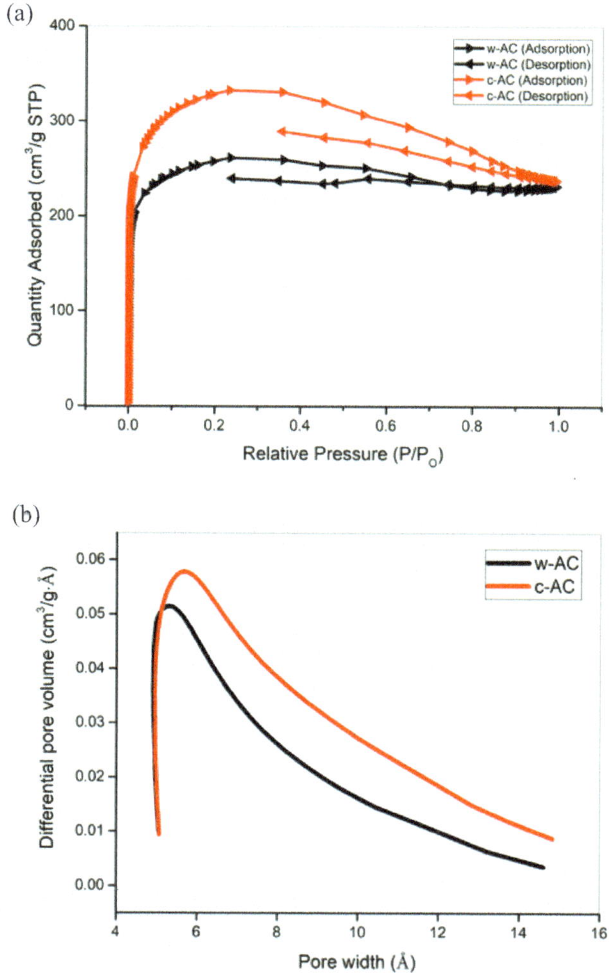

Figure 4.4 (a) BET isotherm and (b) pore size distribution for wood-based and coconut shell-based activated carbons.

According to the IUPAC classification, activated carbons exhibited type I isotherm which is associated with their microporous (< 2 nm) structure with a steep curve at low relative pressure in the range of 0 < P/PO < 0.2. A wide H4 hysteresis loop with a non-reversible desorption isotherm was also detected for the activated carbon due to restricted access to their narrow-slit pores [23]. Coconut shell-based activated carbon exhibited higher adsorption and desorption isotherm than wood-based activated carbon did

Table 4.2 BET surface area and pore volume of the catalysts.

Properties	c-AC	w-AC	Al_2O_3	ZSM-5
BET Surface Area (m^2/g)	1105.89	859.08	2.94	279.94
External Surface Area (m^2/g)	420.59	311.27	4.86	111.88
Micropore area (m^2/g)	685.30	547.81	–	168.06
Pore volume (cm^3/g)	0.45	0.36	–	0.16
Micropore Volume (cm^3/g)	0.33	0.26	–	0.08
Pore size (nm)	1.33	1.66	3.57	2.33

owing to its larger surface area with a more porous structure. Figure 4.4 (b) shows the pore size distribution of the wood-based and coconut shell-based activated carbons where both have a pore diameter range of < 2 nm with a microporosity structure. A sharp increase in pore size distribution at 0.5 nm was detected for the activated carbons. Coconut shell-based activated carbon had a slightly higher differential pore volume than that wood-based activated carbon. Both types of activated carbon with the same pretreatment method exhibited different surface areas and porous volume was due to variation in their sources. A comparison of particle size, pore-volume, surface area, and pore size for the activated carbons is given in Table 4.2.

The wood-based and coconut shell-based activated carbons had a high BET surface area of 859 m^2/g and 1105 m^2/g, respectively. ZSM-5 (279.94 m^2/g) and alumina (2.94 m^2/g), which are commonly used catalysts for the cracking process, had lower surface area than the activated carbons used. Higher liquid yield for activated carbon as a catalyst was justified with its higher BET surface area and porous structure which provided more active sites for the reaction to occur. The catalyst activity of activated carbons sourced from wood and coconut shell was different indeed because of the variation in their physical properties. Coconut shell-based activated carbon with a higher surface area had better catalytic performance than wood-based activated carbon. The pore volume of activated carbons was higher than ZSM-5, while Al_2O_3 had a poorly developed porous structure. A higher microporous area than the external surface area suggests a well-developed porous structure of the activated carbons with more reaction sites. ZSM-5 and Al_2O_3 have a mesoporous structure with a pore size >2 nm. Although the activated carbons possessed a lower average pore size than ZSM-5 and Al_2O_3, they led to lower coke deposition ($\sim$1.2 wt. %). It could be deduced that their microporous configuration did not limit penetration of oleic acid and palmitic acid with the molecular size of 0.516 nm and 0.372 nm [24] to access the active sites within the pores. The high surface area of the activated carbons is suitable for dispersion of active phases [26] to upgrade its catalytic performance. The encouraging cracking and deoxygenation activities towards lighter fuel range hydrocarbons by activated carbon, this had led to further study of using activated carbon-supported metal catalysts for the catalytic cracking reaction.

4.3.2.2 Activated carbon supported metal oxides

To improve the catalyst activity of the activated carbons, the wood-based and coconut shell-based activated carbons were loaded with transition metals. It is believed that those metals converted to oxides during calcination and had been proved via XRD analysis (Section 3.3.2). Table 4.3 shows the liquid, coke, and gas yields obtained from catalytic cracking of waste cooking oil using different activated carbon-supported metal oxide as catalysts. NiO improved the liquid yield from 44–45 wt. % to 54–55 wt. % with a reduced gaseous product for both types of activated carbons. NiO on coconut shell-based activated carbon gave a slightly higher liquid (54.99 wt. %) yield and lower gas yield than wood-based activated carbon. Hence, the coconut shell-based activated carbon was selected to support different metal oxides for further investigation.

In Table 4.3, liquid yields increased in the sequence of CuO $<$ Fe_2O_3 $<$ NiO $<$ MoO_3 $<$ Co_3O_4 supported on activated carbon. The highest liquid yield at 59.02 wt. % was obtained in catalytic cracking using Co_3O_4/c-AC whereas the lowest liquid yield (46.84 wt. %) was obtained by using CuO/c-AC with a high coke yield of 3.78 wt. %. A similar observation was also reported [18]. The low liquid yield by CuO/c-AC was most probably due to the deactivation of the catalyst by coke formation. A high gaseous product exhibited by CuO was also observed by Li et al. [17] due to its high acidity. The lowest coke yield (3.54 wt. %) obtained with the Co_3O_4/c-AC catalyst

Table 4.3 Liquid, coke, and gas yields for catalytic cracking of waste cooking oil using activated carbon-supported metal oxide catalysts at 450 °C; WHSV 9 hr^{-1} and 60 mins.

The catalyst used in cracking reaction	Yield (wt. %)		
	Liquid	Coke	Gas
NiO/w-AC	54.01	3.59	42.40
NiO/c-AC	54.99	3.58	41.43
MoO_3/c-AC	55.81	3.87	40.32
Co_3O_4/c-AC	59.02	3.54	37.44
Fe_2O_3/c-AC	51.85	3.71	44.44
CuO/c-AC	46.84	3.78	49.38

suggested its higher selectivity for liquid products as desired. Interestingly, MoO_3/c-AC gave a high liquid yield of 55.81 wt. % in spite of the highest coke yield (3.87 wt. %). NiO/c-AC had a slightly higher liquid yield (54.99 wt. %) compared to NiO/w-AC which should be caused by the slight difference in properties of the activated carbons which will be discussed in Section 3.3.

Figure 4.5 shows hydrocarbon content in the range of C_5 to C_{20} for the liquid product obtained from catalytic cracking of waste cooking oil with the activated carbon-supported metal oxide catalysts.

Hydrocarbon yield (C_5–C_{20}) in liquid product increased in the order of: MoO_3 (56.13%) < CuO (71.43%) < NiO (76.24%) < Co_3O_4 (79.59%) < Fe_2O_3 (83.27%) on coconut shell-based activated carbon. Although Fe_2O_3/c-AC generated a lower liquid yield (51.85 wt. %) than NiO/c-AC (54.99 wt. %) catalyst, it had the highest liquid C_5–C_{20} hydrocarbon yield (83%) in the study. Loading of NiO, Co_3O_4, and Fe_2O_3 to activated carbon improved the liquid hydrocarbon yield and similar observations had also been reported [19]. However, it was not the case for loading MoO_3 and CuO. MoO_3/c-AC catalyst seemed to favor heavy residual substances with the lowest C_5–C_{20} liquid hydrocarbon yield despite its high liquid product. CuO incorporation to activated carbon had the least liquid yield with a lower C_5–C_{20} hydrocarbon content (73%) which implied its poor catalytic activity in this case.

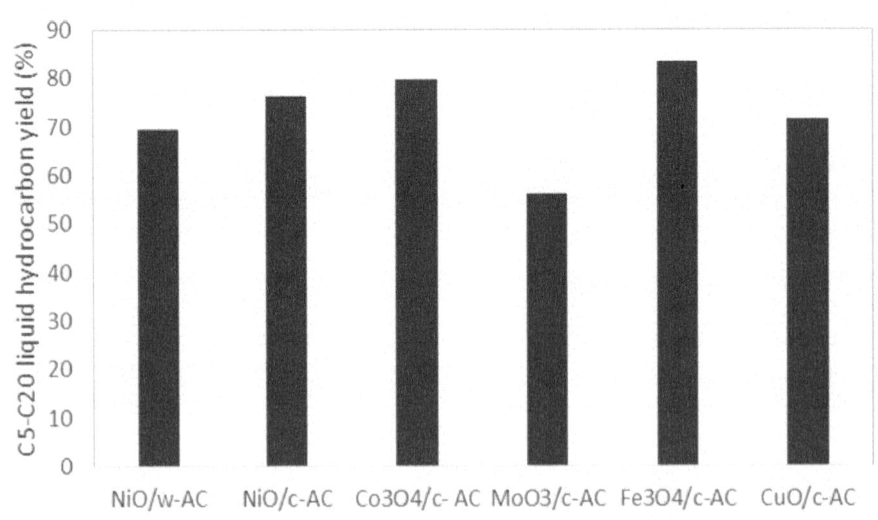

Figure 4.5 Hydrocarbon yield for metal oxide-modified activated carbon in catalytic cracking of waste cooking oil at 450 °C; 9 hr^{-1} and 60 mins.

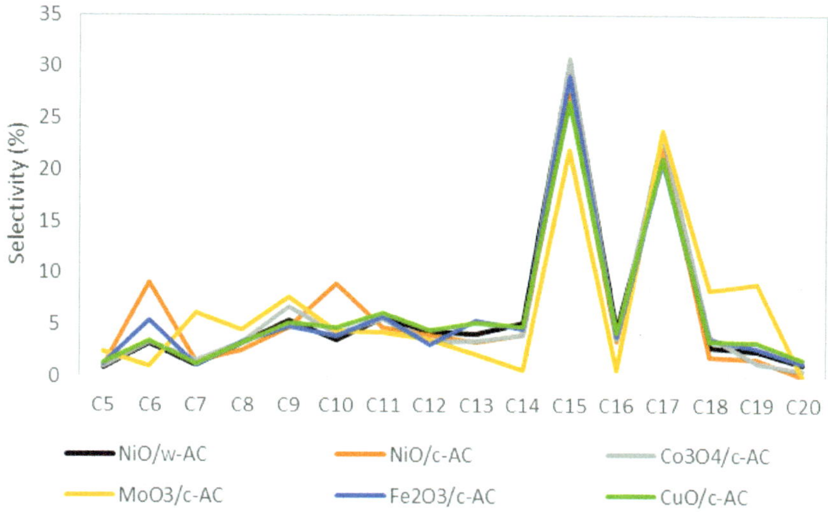

Figure 4.6 Hydrocarbon selectivity of liquid product for catalytic cracking of waste cooking oil using metal oxide modified activated carbon.

Co_3O_4/c-AC exhibited good catalyst activity with a liquid hydrocarbon yield of 79%. Meanwhile, Fe_2O_3/c-AC gave the highest liquid hydrocarbon yield (83.27%) among the studied catalysts which meant more cracking reactions could have been involved.

Figure 4.6 shows the selectivity of liquid product to carbon number C_5 to C_{20} hydrocarbons for the coconut shell-based activated carbon-supported metal oxide catalysts. It was noted that the liquid products consist mainly of C_{15} and C_{17} hydrocarbons which is believed as the consequence of decarboxylation or decarbonylation of triglycerides with mainly fatty acids of C_{16} and C_{18} in waste cooking oil. Despite higher oleic acid (C_{18}) than palmitic acid (C_{16}) in the waste cooking oil, the liquid product was dominated by C_{15} hydrocarbons. This could be due to the deoxygenation of the fatty acids is followed by mild cracking of C_{17} compounds [20]. MoO_3/c-AC favored secondary reactions that produced high heavy C_{18}–C_{20} compounds (17.30%) relatively. In the present study, the highest selectivity towards C_{15} and C_{17} total hydrocarbons (54.36%) was achieved by Co_3O_4/c-AC. The liquid products obtained by Co_3O_4/c-AC and Fe_2O_3/c-AC catalysts with the highest liquid yield as well as selectivity, and C_5–C_{20} liquid hydrocarbon yield, respectively were analyzed for their oxygen content. Their product quality was also compared to that of MoO_3/c-AC with the least C_5–C_{20} liquid hydrocarbon yield and ZSM-5 which gave the minimum liquid product.

Table 4.4 Elemental analysis of the liquid product from catalytic cracking of waste cooking oil (450 °C; WHSV 9 hr^{-1} and 60 mins)

Elements (wt. %)	Waste Cooking Oil	Thermal cracking	Catalytic cracking			
			ZSM-5	MoO_3/ c-AC	Fe_2O_3/ c-AC	Co_3O_4/ c-AC
Carbon	70.85	71.82	73.67	73.83	76.00	76.36
Hydrogen	10.21	11.32	10.52	10.75	11.16	11.35
Nitrogen	0.05	0.26	0.39	0.06	0.05	0.23
Sulfur	0.17	0.09	0.17	0.27	0.24	0.32
Oxygen	18.72	16.51	15.25	15.09	12.55	11.74
Deoxygenation (%)	-	11.81	18.54	19.39	32.96	37.29

Table 4.4 shows the carbon, hydrogen, nitrogen, sulfur, and oxygen content of the selected liquid product from the reactions. The oxygen content of the waste cooking oil initially was around 18.72 wt. % and reduced to less than 15.25 wt. % after catalytic cracking reaction via deoxygenation reaction. Poor deoxygenation during thermal cracking reduced the oxygen content in the waste cooking oil by only 2.21 wt. %. The presence of nitrogen and sulfur detected was caused by oxidation and hydrolysis preservatives found in food [21], which were maintained at a low level (<0.4 wt. %). Weak deoxygenation of waste cooking oil was observed for MoO_3/c-AC catalyst with 19.39% unraveled its lower hydrocarbon yield despite its higher liquid yield relatively. ZSM-5 displayed an even lower efficiency of 18.54% deoxygenation. Deoxygenation of waste cooking oil achieved was 37.29% for Co_3O_4/c-AC catalyst and 32.96% for Fe_2O_3/c-AC catalyst. Generally, Co_3O_4/c-AC and Fe_2O_3/c-AC showed better catalytic cracking and deoxygenation activity with higher liquid yield, and C_5–C_{20} liquid hydrocarbons yield relative to other coconut shell-based activated carbon-supported metal oxides studied. It is believed that Fe_2O_3 induces cleavage of the C–O bond [22] and Co_3O_4 boosts decarboxylation reaction [1].

4.3.3 Characterization of Activated Carbon Supported Metal Catalysts

4.3.3.1 X-ray diffraction (XRD) analysis

The XRD patterns show the phase identity and crystallinity of the catalysts. XRD of the fresh and used coconut shell-based activated carbon-supported

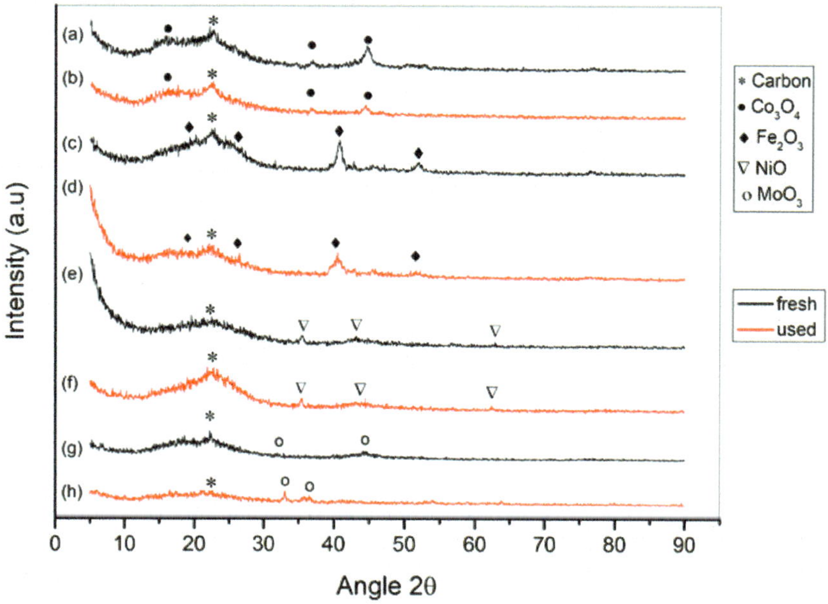

Figure 4.7 XRD patterns for (a) & (b) Co_3O_4/c-AC; (c) & (d) Fe_2O_3/c-AC; (e) & (f) NiO/c-AC and (g) & (h) MoO_3/c-AC.

metal oxides are displayed in Figure 4.7. A major peak observed at $2\theta = 22.5°$ represents the presence of the amorphous phase of carbonaceous substance in all the activated carbon-based catalysts [20].

In Figure 4.7(a) &(b), the peaks for crystal planes of the cubic Co_3O_4 phase at $2\theta = 18.812°$ (1 1 1), 36.474° (3 1 1) and 44.349° (4 0 0) (ICSD Ref. Code 01-080-1537) were distinguished in fresh and used Co_3O_4/c-AC. Co_3O_4 as the active phase is mixed-valence oxides of Co^{2+} and Co^{3+} [28]. The precursor used for cobalt contained Co^{2+} which oxidizes to CoO via reaction with oxygen in the atmosphere. Simultaneously, Co^{2+} was also converted to Co^{3+} to form Co_2O_3. For Fe_2O_3/c-AC catalyst (Figure 3.6(c)–(d)), the peaks at $2\theta = 18.392°$ (1 1 1) and 53.888° (4 2 2) (ICSD Ref. Code 00-004-0755) were due to crystal planes of the cubic Fe_2O_3 phase while peaks at $2\theta = 24.165°$ (0 1 2) and 40.991° (1 1 3) (ICSD Ref. Code 00-001-1053) were due to rhombohedral Fe_2O_3 phase. Both types of Fe_2O_3 crystal shapes were identified in fresh and used Fe_2O_3/c-AC. Fe_2O_3 was formed through the reaction of Fe^{3+} from the ionization of iron precursor with water and oxygen. While, for NiO/c-AC, the peaks for crystal planes of the cubic NiO phase at $2\theta= 37.335°$ (1 1 1), 43.381° (2 0 0), and 63.024° (2 2 0)

(ICSD Ref. Code 01-073-1519) were spotted in fresh and used NiO/c-AC. The peaks for crystal planes of the monoclinic MoO_3 phase at $2\theta = 23.01°$ (0 1 1), 32.161° (0 0 2) and 44.347° (−3 1 1) (ICSD Ref. Code 01-089-1554) were detected for fresh MoO_3/c-AC whereas MoO_3 phases at $2\theta = 33.344°$ (0 2 0), 34.995° (1 0 2) and 54.372° (4 1 0) are noted for used MoO_3/c-AC. Peaks for MoO_3, NiO, Co_3O_4, and Fe_2O_3 implied success in the preparation of the activated carbon-supported metal oxides. Non-distinctive intensities of the peaks were attributed to well-scattered metal oxides with low aggregation of active phases [29] which could lead to rapid catalyst deactivation [26]. Calcination of catalysts at optimum temperature with a slower heating rate could have prevented the growth of crystals for the metal phases [30, 31] and complete oxidation of metal nitrates in the atmospheric air was witnessed too. Peak intensity decreased slightly in used catalysts from the fresh could be due to partial reduction of metal oxides to lower oxidative states in the presence of CO gas emitted during the reaction as reductant. Varied valences of metal oxides dispense oxygen vacancies to store or transfer oxygen [5].

Scherrer's equation is used to estimate the crystallite size of the catalyst:

$$D = \frac{K\lambda}{\beta \cos\theta} \tag{4.6}$$

where D = crystal size; K = shape factor = 0.9; λ = 0.15406 nm; β = FWHM; θ = angle. Table 4.5 shows the crystal size calculated for the activated carbon-based catalysts. Lower crystal size of < 1 nm corresponded to less interaction between active phases and activated carbon support hindered aggregation and crystallization of the homogeneously spread metal oxides. The crystal size of used catalysts reduced from their respective fresh catalysts suggests a reduction of their active phase which explained the drop in its catalytic performance over time. However, crystal size for NiO/c-AC catalyst is retained at 0.08 nm because of the stability of Ni oxide whereas Fe_2O_3 experienced a major drop in crystal size because it could have transformed to another stable oxidation state of FeO.

Table 4.5 Crystal size of catalysts calculated from Scherrer's Equation.

Catalyst	Crystal size (nm)	
	Fresh	Used
MoO_3/c-AC	0.13	0.08
NiO/c-AC	0.08	0.08
Fe_2O_3/c-AC	0.47	0.25
Co_3O_4/c-AC	0.19	0.11

4.3.3.2 Scanning electron microscopy (SEM)

Figure 4.8 shows the SEM images of Mo, Fe, and Co oxides on activated carbon. The porous structure of the smooth-surfaced activated carbon was visible with widely distributed active metal oxides as white spherical particles without agglomeration that allowed good exposure for better catalyst activity. Hole-like structures observed resembles well-developed pores of the activated carbons which were not destroyed by calcination under nitrogen gas flow at steady heating.

On the contrary, calcination in the air would cause oxygen to react with carbon via combustion to release CO_2 and form fly ash [32]. The average pore size for the activated carbon-based catalysts was estimated to be less than 2 μm which validated its microporosity structure as determined in

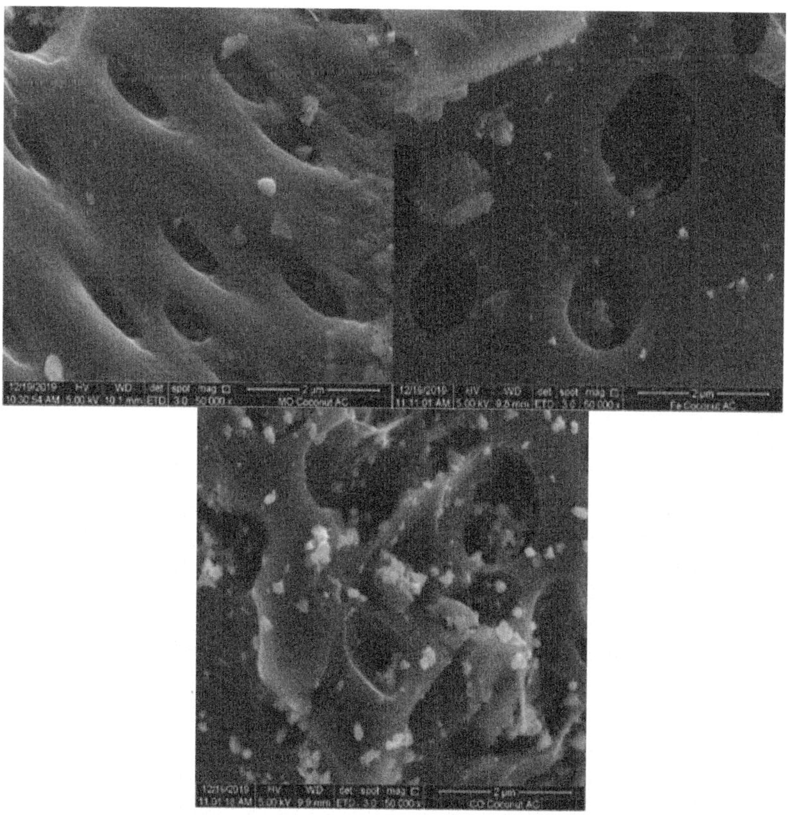

Figure 4.8 SEM images for (a) MoO_3/c-AC; (b) Fe_3O_4/c-AC and (c) Co_3O_4/c-AC catalysts at a magnification of 50000 $\times$.

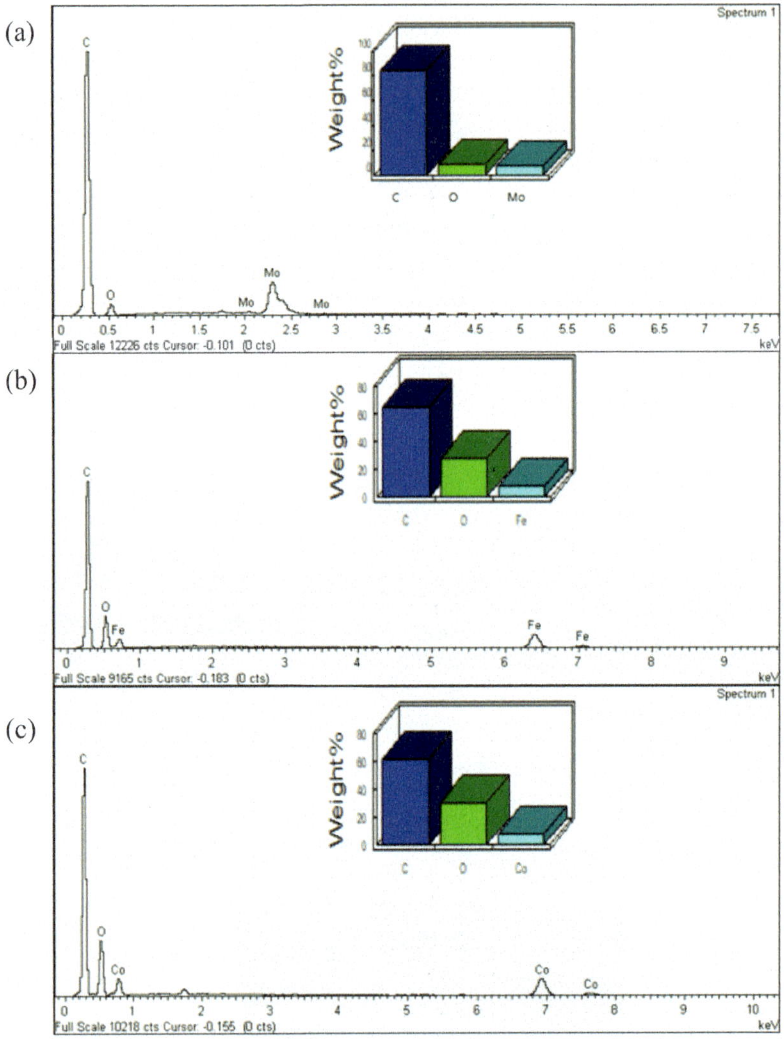

Figure 4.9 EDX spectra for (a) MoO_3/c-AC, (b) Fe_2O_3/c-AC and (c) Co_3O_4/c-AC catalysts.

BET analysis. Figure 4.9 shows the energy-dispersive x-ray spectrum for the activated carbon-based catalysts. MoO_3/c-AC consists of 7.18 wt. % Mo, Fe_2O_3/c-AC contains 7.87 wt. % Fe and Co_3O_4/c-AC hold 8.49 wt. % Co. The metal content was close to the concentration intended to be loaded (10 wt. %) on the activated carbon catalysts. This indicated uniform dispersion of metal impregnated on activated carbon catalyst that limited reduction of its surface area [26] due to accumulation of the active phases which minimizes its contact with reactants. The significant carbon content of more than 60 wt. % of the catalysts was contemplated to provide good mechanical stability and thermal activity. The absence of N in the catalysts is attributed to the complete oxidation of the metal nitrate precursors on activated carbon during calcination [30].

4.3.3.3 Temperature programmed desorption (TPD)

TPD-NH_3 and CO_2 assessed the acidity and basicity of the activated carbon-based catalysts, respectively. Peaks detected at a temperature below 300 °C represent weak acidity or basicity whereas desorption at a temperature beyond 500 °C represents strong acidic or basic sites. Table 4.6 shows the strength and density of acidic and basic sites observed for the activated carbon-based catalysts. Strong acid sites were presented by all the activated carbon-based catalysts which could be contributed by the nitric acid-treated activated carbon. The strong acidity of activated carbon was reduced with the incorporation of metal oxides which provided basic sites and mild acid sites. However, Fe_3O_4/c-AC and MoO_3/c-AC catalysts with only strong acid sites held a higher density of acidic sites than did Co_3O_4/c-AC. Higher coke formation for Mo and Fe oxides suggests the presence of strong acid sites led to polyaromatization as an undesired side reaction [20]. Meanwhile, Fe_3O_4/c-AC exhibited the most basicity with both strong and mild basic sites. But, Co_3O_4/c-AC expressed weak basic sites whereas MoO_3/c-AC possessed medium and strong basic sites. High C_{13} hydrocarbons could be induced by the higher basicity of MoO_3 resulted in increased cleavage of C–C bonds to produce lighter hydrocarbons. More cracking activity was also detected for Fe_3O_4 with more strong basic sites which yielded higher C_7–C_{11} hydrocarbons [33]. Higher density of both acidic and basic sites of Fe_2O_3/c-AC

Table 4.6 Basicity and acidity of catalysts were analyzed using temperature-programmed desorption.

Catalyst	Basicity (cm^3/g STP)			Acidity (cm^3/g STP)		
	Weak	Medium	Strong	Weak	Medium	Strong
c-AC	-	-	-	-	-	34.54
MoO_3/ c-AC	-	14.89	2.70	-	-	21.55
Fe_2O_3/ c-AC	10.41	10.81	8.30	-	-	24.72
Co_3O_4/ c-AC	18.90	-	-	2.34	3.42	12.66

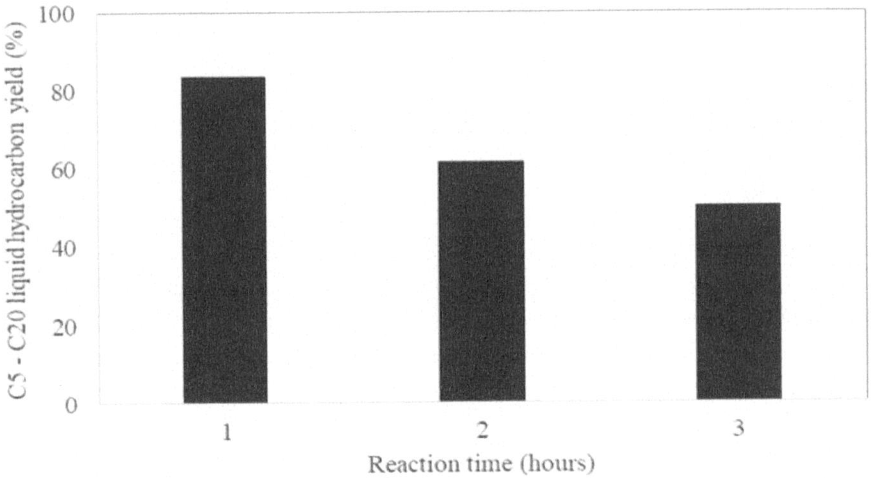

Figure 4.10 Liquid yield and C_5–C_{20} hydrocarbon yield for Fe_2O_3/c-AC in catalytic cracking of waste cooking oil at 450 °C and WHSV 9 hr^{-1} for 3 hrs of reaction.

granted effective deoxygenation and cracking reactions for its maximum C_5–C_{20} hydrocarbon yield. Co_3O_4/c-AC secured ideal mild and strong acidity with weak basic sites to generate higher liquid yield with more C_{15} and C_{17} hydrocarbons through decarboxylation and decarbonylation reaction via C–O scission for its prominent oxygen removal. Higher oxygenates and the lowest hydrocarbon yield was acquired for MoO_3/c-AC which has medium and strong basicity with strong acidic sites.

4.3.3.4 Catalyst stability test

The Fe_2O_3/c-AC with high liquid hydrocarbon yield was chosen as the catalyst for further testing. Its stability performance was tested over 3 hours of continuous reaction and Figure 4.10 shows its C_5–$C2_0$ liquid hydrocarbon yield. The C_5–C_{20} hydrocarbon yield reduced from 83.53% to 50.34% with increasing reaction time. This can be interpreted as rapid deactivation of the activated carbon catalyst used that could have been caused by severe coke deposition.

4.4 Conclusion

Activated carbon showed a remarkable performance with good deoxygenation activity in catalytic cracking reactions. The activated carbon-based catalysts had more liquid yield with C_{15} and C_{17} hydrocarbons produced via decarboxylation or decarbonylation which was higher than ZSM-5 and alumina. The variation in activated carbon source altered its surface area and porous distribution and affected its catalyst activity. Coconut shell-based activated carbon with a higher BET surface area provided more active sites for the reaction. The catalytic activity of activated carbon was improved through loading with transition metal oxides that influenced its acidic and basic properties. The metal oxides (NiO, MoO_3, CuO, Fe_2O_3, and Co_3O_4) supported on activated carbon provided more active sites for reaction and led to higher liquid yield. Co, Fe, and Ni oxides increased C_5–C_{20} hydrocarbon yield to 83%, 79%, and 76%, respectively, in the range of gasoline, jet fuel, kerosene, and diesel fractions with C_{15} and C_{17} hydrocarbons mostly. Better deoxygenation with 37.29% oxygen removal was exhibited by Co_3O_4/c-AC with weak basicity, mild and strong acid sites via decarboxylation, and/or decarbonylation.

Acknowledgment

This study was supported by the Ministry of Higher Education Malaysia for Fundamental Research Grant Scheme with Project Code: FRGS/1/2019/TK02/USM/02/5 (A/C: 203.PJKIMIA. 6071445).

References

[1] Abdulkareem-Alsultan G, Asikin-Mijan N, Mustafa-Alsultan G, Lee HV, Wilson K, Taufiq-Yap YH. Efficient deoxygenation of waste cooking oil over Co3O4–La2O3-doped activated carbon for the production of diesel-like fuel. RSC Adv. 2020; 10: 4996–5009.

[2] Le-Phuc N, Ngo PT, Ha QLM, Tran TV, Phan TT, Luu LC, Duong LT, Phan BMQ. Efficient hydrodeoxygenation of guaiacol and fast-pyrolysis oil from rice straw over PtNiMo/SBA-15 catalyst for co-processing in the fluid catalytic cracking process. J. Environ. Chem. Eng. 2019; 103552.

[3] Srifa A, Faungnawakij K, Itthibenchapong V, Assabumrungrat S. Roles of monometallic catalysts in hydrodeoxygenation of palm oil to green diesel. Chem. Eng. J. 2015; 278: 249–258.

[4] Ameen M, Azizan MT, Yusup S, Ramli A, Yasir M. Catalytic hydrodeoxygenation of triglycerides: An approach to clean diesel fuel production. Renew. Sustain. Energy Rev. 2017; 80: 1072–1088.

[5] Zhang C, Zhang L, Li Q, Wang Y, Liu Q, Wei T, Dong D, Salavati S, Gholizadeh M, Hu X. Catalytic pyrolysis of poplar wood over transition metal oxides: Correlation of catalytic behaviors with physicochemical properties of the oxides, Biomass and Bioenergy. 2019; 124: 125–141.
[6] Zheng Z, Wang J, Wei Y, Liu X, Yu F, Ji J. Effect of La-Fe/Si-MCM-41 catalysts and CaO additive on catalytic cracking of soybean oil for biofuel with low aromatics. J. Anal. Appl. Pyrolysis. 2019; 143: 104693.
[7] Natewong P, Murakami Y, Tani H, Asami K, Bell AT, Gates BC, Ray D, Thompson MR, Gunawan L. Effect of support material on MgO-based catalyst for the production of new hydrocarbon bio-diesel. Am. Sci. Res. J. Eng. Technol. Sci. 2016; 22: 153–165.
[8] Wu J, Shi J, Fu J, Leidl JA, Hou Z, Lu X. Catalytic decarboxylation of fatty acids to aviation fuels over nickel supported on activated carbon. Sci. Rep. 2016; 6: 1–8.
[9] Zhu L, Shen D, Luo KH. A critical review on VOCs adsorption by different porous materials: Species, mechanisms and modification methods. J. Hazard. Mater. 2020; 389: 122102.
[10] Li F, Chen J, Hu X, He F, Bean E, Tsang DCW, Ok YS, Gao B. Applications of carbonaceous adsorbents in the remediation of polycyclic aromatic hydrocarbon-contaminated sediments: A review. J. Clean. Prod. 2020; 255: 120263.
[11] Ooi XY, Gao W, Ong HC, Lee HV, Juan JC, Chen WH, Lee KT. Overview on catalytic deoxygenation for biofuel synthesis using metal oxide supported catalysts. Renew. Sustain. Energy Rev. 2019; 112: 834–852.
[12] Luz GE, Santos AGD, Melo ACR, Oliveira RM, Araujo AS, Fernandes VJ. Thermal catalytic cracking of buriti oil (Mauritia flexuosa L.) over LaSBA-15 mesoporous materials. Fuel Process. Technol. 2011; 92: 2099–2104.
[13] Wang F, Jiang J, Wang K, Zhai Q, Long F, Liu P, Feng J, Xia H, Ye J, Li J, Xu J. Hydrotreatment of lipid model for diesel-like alkane using nitrogen-doped mesoporous carbon-supported molybdenum carbide. Appl. Catal. B Environ. 2019; 242: 150–160.
[14] Collins CD. Implementing phytoremediation of petroleum hydrocarbons. 2007; 99–108. DOI: 10.1007/978-1-59745-098-0_8.
[15] Taufiqurrahmi N, Bhatia S. Catalytic cracking of edible and non-edible oils for the production of biofuels. Energy Environ. Sci. 2011; 4: 1087–1112.

[16] Alsultan GA, Asikin-Mijan N, Lee HV, Albazzaz AS, Taufiq-Yap YH. Deoxygenation of waste cooking to renewable diesel over walnut shell-derived nanorods activated carbon supported CaO-La2O3 catalyst. Energy Convers. Manag. 2017; 151: 311–323.

[17] Li C, Ma J, Xiao Z, Hector SB, Liu R, Zuo S, Xie X, Zhang A, Wu H, Liu Q. Catalytic cracking of Swida wilsoniana oil for hydrocarbon biofuel over Cu-modified ZSM-5 zeolite. Fuel. 2018; 218: 59–66.

[18] Wang S, Cai Q, Chen J, Zhang L, Zhu L, Luo Z. Co-cracking of bio-oil model compound mixtures and ethanol over different metal oxide-modified HZSM-5 catalysts. FUEL. 2015.

[19] Iliopoulou EF, Stefanidis SD, Kalogiannis KG, Delimitis A, Lappas AA, Triantafyllidis KS. Catalytic upgrading of biomass pyrolysis vapors using transition metal-modified ZSM-5 zeolite. Appl. Catal. B Environ. 2012; 127: 281–290.

[20] Asikin-Mijan N, Ooi JM, Abdulkareem-Alsultan G, Lee HV, Mastuli MS, Mansir N, Alharti FA, Alghamdi AA, Taufiq-Yap YH. Free-H2 deoxygenation of Jatropha curcas oil into cleaner diesel-grade biofuel over coconut residue-derived activated carbon catalyst. J. Clean. Prod. 2019; 119381.

[21] Bezergianni S, Voutetakis S, Kalogianni A. Catalytic hydrocracking of fresh and used cooking oil. Ind. Eng. Chem. Res. 2009; 48: 8402–8406.

[22] Asikin-Mijan N, Rosman NA, Abdulkareem-Alsultan G, Mastuli MS, Lee HV, Nabihah-Fauzi N, Lokman IM, Alharthi FA, Alghamdi AA, Aisyahi AA, Taufiq-Yap YH. Production of renewable diesel from Jatropha curcas oil via pyrolytic-deoxygenation over various multi-wall carbon nanotube-based catalysts. Process Saf. Environ. Prot. 2020; 142: 336–349.

[23] Gargiulo V, Gomis-Berenguer A, Giudicianni P, Ania CO, Ragucci R, Alfè M. Assessing the potential of biochars prepared by steam-assisted slow pyrolysis for CO2 adsorption and separation. Energy and Fuels. 2018; 32: 10218–10227. https://doi.org/10.1021/acs.energyfuels.8b01058.

[24] Guo W, Zhu Y, Han Y, Luo B, Wei Y. Separation mechanism of fatty acids from waste cooking oil and its flotation performance in iron ore desiliconization . Minerals. 2017; 7: 244.

[25] Parrillo F, Ruoppolo G, Arena U. The role of activated carbon size in the catalytic cracking of naphthalene. Energy. 2020; 190: 116385.

[26] Dehghani S, Haghighi M. Sono-enhanced dispersion of CaO over Zr-Doped MCM-41 bifunctional nanocatalyst with various Si/Zr ratios for

conversion of waste cooking oil to biodiesel. Renew. Energy. 2020; 153: 801–812. https://doi.org/10.1016/j.renene.2020.02.023.

[27] Jeon KW, Shim JO, Jang WJ, Lee DW, Na HS, Kim HM, Lee YL, Yoo SY, Roh HS, Jeon BH, Bae JW, Ko CH. Effect of calcination temperature on the association between free NiO species and catalytic activity of Ni-Ce0.6Zr0.4O2 deoxygenation catalysts for biodiesel production. Renew. Energy. 2019; 131: 144–151.

[28] Yigezu ZD, Muthukumar K. Catalytic cracking of vegetable oil with metal oxides for biofuel production. Energy Convers. Manag. 2014; 84: 326–333.

[29] Bayat A, Sadrameli SM Production of renewable aromatic hydrocarbons via conversion of canola oil methyl ester (CME) over zinc promoted HZSM-5 catalysts. Renew. Energy. 2017; 106: 62–67.

[30] Alaei S, Haghighi M, Rahmanivahid B, Shokrani R, Naghavi H. Conventional vs. hybrid methods for dispersion of MgO over magnetic Mg–Fe mixed oxides nanocatalyst in biofuel production from vegetable oil. Renew. Energy. 2020; 154: 1188–1203.

[31] Mohammadpour M, Haghighi M, Shokrani R. Influence of heating approach (microwave vs. muffle furnace) and fuel in auto-combustion design of nanostructured Ca2Mn3O8 as support for efficient and reusable catalyst used in green fuel production. Ceram. Int. 2020; 46: 12180–12191.

[32] Abdulkareem-Alsultan G, Asikin-Mijan N, Lee HV, Rashid U, Islam A, Taufiq-Yap YH. Review on thermal conversion of plant oil (Edible and inedible) into green fuel using carbon-based nanocatalyst. Catalysts. 2019; 9: 350. https://doi.org/10.3390/catal9040350.

[33] Przekop RE, Osinska-Broniarz M, Martyła A, Sztorch B, Stodolny M, Kirszensztejn P, Szymański M. Catalytic cracking of rapeseed oil with binary oxide systems: an alternative to production of petrochemicals, JAOCS. J. Am. Oil Chem. Soc. 2020.

5

Biofuels – Are they a Sustainable Alternative?

P. S. Ranjit[1], Swapnil S. Bhurat[2], G. S. Mahesh[3], Sreenivasa Reddy[1], Yashvir Singh[4,5], and Wei-Hsin Chen[6,7,8]

[1]Department of Mechanical Engineering, Aditya Engineering College, Surampalem, India
[2]University of Petroleum & Energy Studies (UPES), Dehradun, India
[3]Geetha Shishu Shikshana Sangha Institute of Engineering and Technology for Women, Mysuru, India
[4]Department of Mechanical Engineering, Graphic Era Deemed to be University, Dehradun, Uttarakhand, India
[5]Faculty of Mechanical and Manufacturing Engineering, Universiti Tun Hussein Onn Malaysia, Parit Raja, Batu Pahat, Johor, Malaysia
[6]Department of Aeronautics and Astronautics, National Cheng Kung University, Tainan, 701, Taiwan
[7]Research Centre for Smart Sustainable Circular Economy, Tunghai, 407, Taiwan
[8]Department of Mechanical Engineering, National Chin-Yi University of Technology, Taichung, 411, Taiwan
E-mail: psranjit1234@gmail.com; yashvirsingh21@gmail.com

Abstract

Biofuel, being an organic material, is extracted from natural resources; its utilization in the form of Land, Water, and Solar energy was discussed. Moreover, its impact on Food and its associated areas was revealed. Further, the energy content of the biodiesel, along with solar energy utilization in framing the biomass, was in detail explored.

Keywords: Biofuel, Land, Water, Solar, Energy content.

5.1 Introduction

Worldwide deficiencies of fossil energy, particularly oil and gaseous petrol, have expanded interest in biomass and biofuels worldwide [1]. This accentuation on biofuels as sustainable power sources has grown worldwide, particularly those biofuels produced using yields like edible oils like corn, sugarcane, canola, assault, and soybean [2]. Even though it might be advantageous to utilize sustainable plant materials for biofuel, the utilization of yields and other biomass materials for biofuels raises numerous ecological and moral concerns [3]. Various contentions exist in utilizing land, water, energy, and other natural assets for Food and biofuel creation.

5.2 Abstraction of Biofuels from Food

India is the second largest agricultural land after the United States; its utilization in India is 20.3% higher [4] than U.S. with 40% [2] for food production. Water and Food are essential to people and any remaining creatures, plants, and organisms on Earth. Per capita food supplies (cereal grains) for people have been declining for around 20 years, to some extent due to deficiencies of freshwater and cropland and the simultaneous expansion in human numbers. Deficiencies in food supplies have added to the increment in the number of malnourished individuals on the planet to more than 4.4 billion, 66% of the populace [5]. Two of the most genuine lack of healthy sustenance issues include iron insufficiency, which influences 2 billion individuals, and protein/calorie inadequacy, which influences 925 million [5, 6]. The iron and protein/calorie inadequacies each outcome in around ten lakh passing per year. People acquire practically the entirety of their supplements from harvests and domesticated animals, and these supplement sources require water, land, and energy for creation. Food and biofuels are subject to these equivalent assets for creation. Various struggles exist in utilizing land, water, energy, and other natural assets for Food and biofuel creation. The world's total populace at present numbers 7 billion, with more than a fourth of 1,000,000 individuals being added every day [7]. The total populace is projected to reach around 13 billion individuals by 2070, dependent on a pace of increment of 1.2% each year.

Moreover, freshwater requests worldwide have been expanding quickly as the populace and economies develop [8]. Populace development, joined by expanded water use, will not just seriously diminish water accessibility per individual but also stress all parts of the whole worldwide environment [9].

Central points impact water accessibility, including precipitation, temperature, dissipation rates, soil quality, vegetation type, and water spillover. Furthermore, genuine challenges exist in reasonably assigning the world's freshwater assets between and inside nations. Generally speaking, water deficiencies seriously decrease food creation; contrarily sway both earthly and amphibian conditions; lessen plant, creature, and organism biodiversity; work with the spread of genuine human infections, and reduce water quality.

5.2.1 Water Resources

Of the assessed 49.44 x 10^{11} TMC of water on the Earth, over 97% is in the seas. Around 12.36 x 10^{8} TMC, or just 3% of Earth's water, is freshwater, of which about 0.3% is held in streams, lakes, and repositories [10]. The rest of the freshwater is put away in glacial masses, perpetual snow, and groundwater springs. The Earth's environment contains around 460 TMC of water and is the wellspring of all the downpour that falls on Earth. Yearly, around 151,000 quads (1 quad = 1015 BTU) of sunlight-based energy cause vanishing and move around 20.37 x 10^{3} (0.04%) of the water from the Earth's surface into the climate. Of this dissipation, 86% is from seas. Albeit just 14% of the water, dissipation is from land, about 20% (4061 TMC each time) of the world's precipitation falls ashore with the excess water getting back to the seas through waterways. In this manner, Sun-powered energy moves 6% of the water from the seas to land regions every year. This part of the hydrologic cycle is crucial not exclusively to horticulture yet additionally to human existence and normal environments [11]

5.2.1.1 Availability of water

Although water is viewed as a sustainable asset since it is renewed by precipitation, its accessibility is limited as far as the sum accessible per unit time in any one locale. The normal precipitation for most landmasses is around 700 mm/yr, yet changes among and inside them. As a general rule, a country is viewed as water scant when the accessibility of water dips under 1,000 m3/yr per capita [12]. Regardless of having a normal of 640 mm/yr of precipitation, Africa is somewhat dry since its high temperatures and winds cultivate quick dissipation [13]. World locales with low precipitation (under 500 mm/yr) experience genuine water deficiencies and lack harvest yields. For instance, 9 of the 14 Middle Eastern nations have scarce freshwater [14]. Considerable withdrawals from lakes, streams, groundwater, and repositories are utilized to address people's issues, urban areas, ranches, and businesses, and these withdrawals stress the accessibility of water in certain pieces of the United States [15]. The all-out agrarian, cultural, and natural framework should be thought of when overseeing water assets. Enactment is some of the time needed to guarantee a reasonable allotment of water. Groundwater assets Approximately 30% (38 x 10^4 TMC) of all freshwater on Earth has been put away as groundwater. The water measured as groundwater is more than multiple times the aggregate sum gathered in waterways and lakes. Most groundwater has gathered for more than a long period in immense springs situated underneath the outside of the Earth. Springs are renewed gradually by precipitation, with a normal re-energize rate that reaches from 0.1% to 3% each year [8, 16, 17]. Accepting a normal 1% re-energize rate, just 3884 TMC of water each year is accessible for practical utilization around the world. As of now, world groundwater springs give around 23% of all water utilized worldwide. In India, as of April 2015, the country's water asset potential or yearly water accessibility as far as regular overflow (stream) in waterways is around 66 x 103 TMC/year. However, the usable water assets of the nation have been assessed as 39.65 x 103 TMC/year. Out of this 39.65 x 103 TMC/year 24. 36 x 103 TMC/year is available as a surface eater and the rest as groundwater [18].

Populace development, expanded flooded farming, and other water utilizes are mining groundwater assets. In particular, the uncontrolled pace of water withdrawal from springs is fundamentally quicker than the regular pace of re-energize, causing water tables to fall by more than 30 m in some U.S. areas somewhere between 1950 and 1990 [19]. The best paces of exhaustion happened in the planet's biggest farming areas, including India, north-eastern China, and California's Central Valley [20]. Comparative issues exist all through the world. For example, in the agronomically useful Chenaran Plain in north-eastern Iran, the water table has declined by 2.8 m each year since the last part of the 1990s [21].

5.2.1.2 Stored water assets

Around 8000 dams are available globally to store the water. Out of those, 57000 are large size dams. India holds 5102 rankings after America and China. The normal existence is around 50 years, and 85% of U.S. dams will be over 50 years of age by 2020 [21]. Possibilities for the development of new dams in the United States do not seem to empower. Moreover, after some time, the limit of all dams is decreased as residue aggregates behind them.

5.3 Water Usage

Water from various sources is removed both for use and utilization in different human exercises. The term used alludes to all human exercises for which a portion of the removed water is returned for reuse, like cooking water, wash water, and waste water. Interestingly, utilization implies that the removed water is nonrecoverable. For instance, the evapotranspiration of water by plants is delivered into the climate and is considered nonrecoverable. To support wellbeing, people should drink from 1.5 to 2.5 liters of water per individual each day [22]. As well as Americans use around 1,410 liters/individual/day of water for cooking, washing, discarding squanders, drinking water, and other individual uses [23]. Whereas Indians use 135 lpcd for domestic purposes like gardening, flushing, cooking, bathing, washing utensils, clothes, etc... [24]. Globally, the average extraction volume for all uses is 1,970 litres/person/day [25]. Approximately 70% of the world's extracted water is consumed and cannot be recycled

5.3.1 Usage of Water in the Growing Crop

Plants need water for photosynthesis, growth, and reproduction. The water used by plants cannot be recovered because some water is chemically released into the atmosphere, part of the plant. In fixing carbon dioxide and controlling the temperature, plants need to emit a huge amount of water. Various crops produce water at 600–2,000 liters per kilogram of desiccant in the produced crop. It is estimated that the average global movement of water into the atmosphere in terrestrial ecosystems due to increased vegetation is about 64% of all precipitation on Earth [26]. The minimum soil moisture essential for crop growth varies. For example, it has been reported that American potatoes require 25–50%, and maize 50–70% [27], whereas Indian and Chinese rice require at least 80% soil moisture. In the semi-dry state, the yield of unirrigated crops such as corn is low even with sufficient fertilizers (1–2.5 tons/hectare). In the tropics, one hectare needs 12 to 16 million liters of water to produce 85 tonnes of sugar cane or 1500 liters of water to produce 1 kg of sugar [28]. In central, south-eastern Brazil, it is estimated that 150 m3 of water is used to produce 1 ton of sugar cane. World farming consumes roughly 70% of freshwater removed each year [29]. Around 17% of the world's cropland is flooded, yet that land produces 40% of the world's Food [30]. Around the world, the measure of flooded land is gradually growing, although salinization, waterlogging, and siltation keep diminishing their efficiency [25]. Regardless of a little yearly expansion altogether watered regions, the per capita flooded region has been declining since 1990 because of fast populace development. In particular, the worldwide water system per capita has declined by almost 10% from 2000 to 2010.

5.4 Biofuels and their Energy Content [31]

Biofuels are fuels that are received from natural substances. They consist of, however, are not confined to, wooden, straw, a sugarcane residue, animal wastes, landfill fuel, biodiesel and ethanol, municipal waste, black liquor (a power-wealthy liquid produced through the pulp and paper industry), switchgrass, and a lot of artificial gases derived from plant and animal be counted. Biofuels are wonderful from fossil fuels in that they are, in a few senses, renewable. There are so many biofuels, and they are derived from many sources that their description calls for a type scheme. In truth, biofuels are usually categorized in numerous approaches, 3 of which might be defined here. The first type of scheme makes use of 3 classes: (1) wooden fuels, (2) agricultural fuels, and (3) city waste-primarily based fuels. In addition to gasoline wood, the class "wooden fuels" consists of charcoal, which is synthetic through partly combusting wood in a low-oxygen environment, and black liquor. Black liquor, a powerful, wealthy liquid derived from wood, is burned through the organizations that produce it, in component to reap the power, it had to keep production operations and get better non-combustible substances suspended inside it. Wood can also be used as a feedstock or uncooked cloth to fabricate a flammable fuel via a system known as gasification or to provide ethanol, even though neither the gasification system nor the wooden-to-ethanol conversion system has reached the factor wherein the generation is prepared to be commercialized. There is, however, an awesome deal of hobby in each procedure. Gaseous and liquid fuels are regularly favored to direct combustion of strong plants be counted because they tend to burn greater cleanly and may be utilized in approaches that strong gasoline cannot. Wood fuels are, then, as easy as logs or bark fed without delay right into a furnace or the result of state-of-the-art processing technologies.

Agricultural fuels normally fall into one every class: mainly grown plants and agricultural waste. In the mainly grown class are gasoline plants together with switchgrass, which may be burned without delay to provide warmness or, in theory, processed to provide ethanol or a flammable fuel. Currently, the transportation fuels ethanol and biodiesel are received from many agricultural plants—ethanol is regularly constituted of corn or sugarcane, and biodiesel is constituted from many styles of plant life and grease from eating places meal-processing organizations. In the rural waste, the class is animal wastes, corn stover, a part of the plant left in the back after the corn kernels were harvested, and bagasse, the part of the sugarcane plant final after the juice has been removed.

Rice husks, a spinoff of cultivating one of the world's maximum extensively grown meal plants, are any other instance of agricultural waste that may be transformed into power. Finally, city waste-primarily based fuels include diverse kinds of municipal waste, consisting of creation debris, furniture, backyard waste (together with leaves and grass cuttings), antique tires, and the methane produced in landfills. The capacity contribution that may be made through municipal waste is small in the evaluation of the capacity of the alternative classes; however, there are, as may be seen, proper motives to apply it.

Another beneficial and unusual place for classifying biofuels is the handiest classes, present-day and traditional. Ethanol, biodiesel, artificial fuel, and black liquor are examples of present-day fuels because they are synthetic fuels or, withinside the case of black liquor, the spinoff of a production system. Conventional fuels consist of unprocessed wood, charcoal, and straw. The use of traditional fuels extends away again in records; however, they stay important. Charcoal, for instance, remains the primary power supply for 200–three hundred million people. Many fuels span the present day and traditional classes, and manure is one of those. Nevertheless, manure is likewise present-day gasoline because a few farms and feedlots system manure to provide methane fuel that's burned for warmth or electricity generation.

A 1/3 not unusual place and beneficial scheme classify kinds of biomass, the uncooked cloth from which fuels are created, through the volume to which the substances were modified from their unique state.

This scheme is much like the traditional/present-day type scheme; however, it uses three classes. There are number one assets, consisting of gasoline wood, crop residues together with the stalks and leaves left in the back after harvest, grains together with corn, and mainly grown grasses. Secondary assets consist of synthetic fuels, ethanol and biodiesel, and the through-merchandise of producing procedures with black liquor and sawdust. Tertiary assets consist of the styles of fuels that might be generated in greater city environments, and that might pose disposal issues have been they are now no longer burned and transformed into power. Tertiary assets consist of the creation and demolition of debris, municipal wastes, and landfill gases.

Multiple type schemes replicate the truth that each manufacturing and biofuels have severe ramifications for the society that enlarge a ways past the power sector.

These troubles consist of (however are not confined to) the environment, the value of meals, and ethics whether, for instance, it is far moral to apply agricultural assets to provide motor-automobile gasoline while its manufacturing reasons meals charges increase. In short, is it moral to burn one's meals? The troubles related to the manufacturing and use of biofuels tend to be more complicated than the ones related to different power sources. However, before addressing any of those troubles, it is far really well worth searching at an early record of hearthplace and gasoline. Biofuels are valued due to the fact they are reasserting thermal strength. Sometimes this strength is used to warm a domestic or business, and on occasion, it is miles used as gas for a warmth engine—a machine, which includes a vehicle or electric producing station—that converts thermal strength into work. In any case, all different matters are equal; the greater thermal strength a unit of gas produces while burned, the greater precious that gas is. This increases the query of ways plenty of strength is launched by burning a unit of a selected gas.

Before the strength content material of biofuel is measured, it is miles crucial to be specific approximately what is being burned. Roughly speaking, the quantity of thermal strength launched while burning a gas pattern relies upon three crucial characteristics, the primary of which is the gas's"gas issue." The different essential additives are water content material and ash content material. (Ash is the non-flammable stable issue.) The gas issue of many nonwoody samples of biomass—substances as distinctive as corn cobs and rice husks—launch kind of the equal quantities of thermal strength in keeping with unit mass furnished the water and ash have first been removed.

Nevertheless, the water content material of a gas supply can range broadly from pattern to pattern, even amongst samples made out of equal species of plant. Water content material has a huge impact on the to be had thermal power while a pattern of biomass is burned. For instance, timber that consists of approximately 15percentage of water, while measured on a unit-mass basis, will launch approximately two times as an awful lot of warmness as inexperienced timber, which includes approximately 50 per cent water. Consequently, while discussing biofuels, engineers are constantly cautious about becoming aware of the water content of the pattern they have in mind. "Dry" gasoline, additionally referred to as air dry, consists of approximately 15 per cent moisture; due to the fact that left withinside the air to dry, moisture content material will now no longer usually drop under 15 percentage. Fuel that has had all the water eliminated is frequently referred to as "bone dry" or oven dry. Removing all the water from biomass delivery is power-intensive, and it is not generally performed in exercise. Most frequently, air dry ought to suffice.

The ash content material of biomass also can range widely. To go back to the formerly noted instance of corncobs and rice husks, corncobs comprise approximately one percentage ash—this is, the simplest one percentage of a corncob, while measured via way of means of weight, includes non-flammable solids. By contrast, approximately 15 per cent of a rice husk is ash. Therefore, while the same weights of corncobs and rice husks are burned, greater warmness may be generated from the corncobs than the rice husks because much less of the rice husk is flammable—this is to say, a great fraction of dry rice husk is not gasoline at all. In fact, in a pattern of air dry rice husks, the simplest 70 percentage of the pattern is flammable; the last 30 percentage includes water and ash.

Ash, in addition, complicates using any gasoline because, from an environmental factor view, ash can constitute a disposal problem. Therefore, on a keeping with-mass basis, air dry rice husks are, compared to corncobs, power-bad and grimy gasoline.

However, if the previous comments assist in becoming aware of why a few biofuels generate greater thermal power than others, it says nothing in absolute phrases approximately the quantity of thermal power a given gasoline releases while burning. The power content material of selected gasoline is generally expressed as the quantity of thermal power launched in keeping with unit mass or phrases of thermal power launched in keeping with unit quantity. Many technological know-how books emphasize the significance of measuring matters on a keeping with-mass basis, due to the fact mass is greater "fundamental"withinside the feel that the mass of a pattern of the fabric will now no longer extrude if the fabric adjustments area or segment (from the Earth to the Moon or from a strong to a liquid, for instance). Nevertheless, regardless of the price of "in keeping with-mass" measurements from a systematic perspective, engineers and scientists withinside the discipline of transportation fuels frequently select to speak approximately power content material in keeping with unit quantity.

Although mass can be greater fundamental, quantity is frequently greater relevant. The quantity of area in an automobile or aircraft dedicated to sporting gasoline could be very limited, so the quantity (now no longer the mass) of the gasoline is the crucial characteristic. Gasoline that calls for a huge garage quantity could now no longer be sensible to apply withinside the transportation sector due to barriers to the quantity of area that may be dedicated to the garage. By manner of instance, the power content material of ethanol is generally expressed as 75, seven hundred Btu in keeping with a gallon (21.1 MJ/l), and the power of fuel is ready 116,000 Btu in keeping with a gallon (32.2 MJ/l). An automobile that ran on natural ethanol could, therefore, require a gasoline tank that changed into 50 percentage large than that of an in any other case, the same automobile that ran on natural fuel is a good way to have an identical cruising range, and a bigger tank way much less garage and passenger room.

The power content material of gaseous fuels is generally expressed in phrases of the quantity of thermal power launched while burning a unit quantity of fuel at a "well-known" temperature and pressure, which in this book will suggest 60 °F (16.3 °C) and one ecosystem pressure. The power content material of herbal fuel, for instance, is ready 930 Btu in keeping with cubic feet (34.6 MJ/m^3) at well-known temperature and pressure. Different structures of size are used for extraordinary styles of gasoline, a good way to better replicate the bodily residences of the gasoline in the query and the situations beneath neath which it is miles saved and used. Finally, it is miles critical to realize that there are no unusual approaches to expressing power content material. Sometimes the power content material of gasoline is expressed in phrases of the higher heating value (HHV), occasionally in phrases of the lower heating value (L.H.V.). The distinction between the two heating values relies upon the size is made.

Burning any biofuel produces thermal power, carbon dioxide (CO_2), and water in most cases. (This water is similar to the water that changed into already gift withinside the gasoline previous to combustion.) The water created at some stage in the combustion response starts with the shape of water vapor. As the recent combustion gases increase and cool, a number of that water vapor might also condense to shape liquid water. As the water adjustments segment from a vapor to a liquid, a good buy of power is

launched into the encircling medium without generating a extrude withinside the temperature of the water. One can degree the thermal power produced via the combustion response earlier than the water adjustments segment or after. If one measures the thermal power earlier than the vapor-to-liquid segment extrude, one receives the L.H.V. This is not an unusual exercise for transportation fuels because warmness engines usually do now no longer convert the power launched at some stage in the segment extrude into work. However, occasionally, the size is not made at the identical temperature because of the L.H.V. size, however, after the segment extrudes have occurred.

This size is probably made while the gasoline is burned surely for the thermal power it releases—in a residential furnace, for instance—and now no longer for the quantity of thermal power that may be transformed to work. If the size of the thermal power produced is made after the vapor-to-liquid segment extrudes, one obtains the HHV [31].

5.5 Is Biomass is a form of Solar Energy [31]

Biomass is a shape of sun power within side the feel that vegetation converts sun power into chemical power, and biomass solves the trouble of "intermittency" within side the delivery of power from the Sun. At any given area on Earth's surface, the delivery of solar power is unsteady.

Consequently, it is not always constantly to be had while it is miles needed. No sun power is to be had at night, for example, and the fee at which solar power is furnished is tons decreased on cloudy days.

The purpose that intermittency is a primary barrier for those searching to make direct use of solar power is that there is no way to shop for big quantities of strength. One can produce strength by using traditional sun strength technology for so long as the Sun shines; however, while the Sun ceases to shine, there is regularly little to do, wait till the Sun shines again.

Biomass is an exception. Because solar power is saved as chemical power, it may be launched while the biomass is burned, and biomass may be burned for the user's benefit. Because biomass solves the trouble of intermittency, the following step in figuring out its cost is to estimate how effectively it shops sun power this is, while the Sun shines on an area of vegetation, how tons of its power is saved withinside the tissue of the vegetation as chemical power?

Mathematically, a hard estimate of biomass's performance as a garage medium is easy to obtain. Each step withinside the conversion manner may be expressed as more than a few representing the conversion performance of that step. The numbers are then elevated collectively to provide an estimate of the performance of the general manner. The following "performance equations" are estimates of the share of the Sun's power that can, below the fine of circumstances, be transformed into biomass. First, however, it is miles vital to study a few data approximately how inexperienced vegetation converts sun power into chemical power.

The mechanism via how vegetation converts daylight into chemical power is referred to as photosynthesis. Sunlight offers the power to strengthen the photosynthetic reaction, which, more or less speaking, includes changing carbon dioxide and water into glucose and oxygen. Therefore, a precise location to begin in any description of this phenomenon is with the character of daylight.

Visible mold is a shape of electromagnetic power, and as such, it stocks many homes in a not unusual place with different kinds of electromagnetic power, consisting of X-rays and radio waves. In particular, all kinds of electromagnetic power tour in waves. Scientists perceive every sort of electromagnetic power via its wavelength: the space from the height of 1 wave to the height of the following.

For seen mild, the space from top to top is measured in millionths of a meter. Different wavelengths of seen mild may be prominent via way of means of the human eye because unique wavelengths are perceived as unique colors. (Visible mild occupies handiest the small part of the spectrum of all electromagnetic waves, and electromagnetic waves with wavelengths out of doors the slender area of the electromagnetic spectrum containing seen mild cannot be detected via way of means of the eye.) When all colors (or wavelengths) of seen mild are blended collectively, the result is ordinary (white) daylight. The unique wavelengths of seen mild also can be unmixed. A rainbow, for example, will separate the colors of which daylight is composed, making every formerly invisible color seen to the observer. In a rainbow, the shortest wavelengths seem closer to the interior of the rainbow.

The farther to the out of doors of the rainbow one looks, the longer the wavelengths one perceives. The bands of color in a rainbow include growing wavelengths, violet, indigo, blue, inexperienced, yellow, orange, and red. There are different electromagnetic rays found in daylight that are not seen by the eye. Ultraviolet mild, for example, has a wavelength this is shorter than violet, and infrared mild has a wavelength this is longer than red. However, those electromagnetic waves have many homes in a not unusual place with seen mild waves; neither infrared nor ultraviolet electromagnetic waves will cause the photosynthetic reaction.

5.6 Conclusion

This chapter reveals the formation of biomass with the help of land, water, solar energy, and associated ways. The use of edible resources is not yet a viable option in getting biofuels. In India, especially Jatropha, Polanga, and

Karanja, such biofuels are preferred under the "National policy on Biofuels" to avoid the food crunch by using edible feedstock. For agriculture purposes, especially in villages, for short-run applications, the use of non-edible biofuels either in the form of straight vegetable oils or its derived biodiesel will help improve the local people's financial growth and lower the dependency on fossil fuels.

References

[1] Barbara, J. The false promise of biofuels. in Special report from the international forum on globalization and the institute for policy studies. 2007.

[2] Pimentel, D., Renewable and solar energy technologies: energy and environmental issues, in Biofuels, Solar and Wind as Renewable Energy Systems. 2008, Springer. p. 1-17.

[3] Pimentel, D., et al., Conservation of biological diversity in agricultural, forestry, and marine systems, in Food, Energy, and Society, Third Edition. 2007, C.R.C. Press. p. 221-244.

[4] Kaul, V., https://www.firstpost.com/business/india-enough-land-farming-bigger-issues-worry-2032327.html.2015.

[5] Organization, W. H., Nutrition for health and development: a global agenda for combating malnutrition. 2000, World Health Organization.

[6] F. A. O., Crops, and drops: Making the best use of water for agriculture. 2009.

[7] Haub, C. and M. Yanagishita, World Population Data Sheet. Washington. DC: Population Reference Bureau, 1995.

[8] Gleick, P. H., Water in crisis. Vol. 100. 1993: New York: Oxford University Press.

[9] Vörösmarty, C. J., et al., Global water resources: vulnerability from climate change and population growth. science, 2000. 289(5477): p. 284-288.

[10] Shiklomanov, I. A. and J. C. Rodda, World water resources at the beginning of the twenty-first century. 2004: Cambridge University Press.

[11] Jackson, R. B., et al., water in a changing world. Ecological Applications, 2001. 11(4): p. 1027-1045.

[12] Engelman, R. and P. LeRoy, Conserving land: Population and sustainable food production. 1995.

[13] Ashton, P. J., Avoiding conflicts over Africa's water resources. Ambio, 2002: p. 236-242.
[14] Myers, N., Perverse subsidies: how tax dollars can undercut the environment and the economy. 2001: Island Press.
[15] Alley, W. M., T. E. Reilly, and O. L. Franke, Sustainability of groundwater resources. Circular 1186. U.S. Geological Survey, Reston, Virginia, U.S.A., 1999.
[16] Covich, A. P., Water and ecosystems. Water in crisis, 1993: p. 40-55.
[17] La Salle, C. L. G., et al., Renewal rate estimation of groundwater based on radioactive tracers (3H, 14C) in an unconfined aquifer in a semi-arid area, Iullemeden Basin, Niger. Journal of Hydrology, 2001. 254(1-4): p. 145-156.
[18] Suhag, R., 2016. P. R. S., Overview of Ground Water in India.
[19] Brown, L., Rising temperatures and falling water tables rising food prices record world grain harvest shortfall of 83 million tons in prospect. Earth Policy Institute Web site, 2002. 21.
[20] Wada, Y., et al., Global depletion of groundwater resources. Geophysical research letters, 2010. 37(20).
[21] Andreoli, C. and D. Pimentel, Water, Food, and biofuels. Global Economic and Environmental Aspects of Biofuels, 2012: p. 249.
[22] FOODS, I. O. N. Q. O., SPECIAL REPORTS.
[23] Melroy, S. P., Demographic Shifts in the United States Air Force: 1994-2007. 2008, AIR WAR COLL MAXWELL AFB AL.
[24] Grover, A. R., Evaluation of Norms and Standards for Domestic Water Use in Indian cities. American Journal of Sustainable Cities and Society. 1: p. 44-55.
[25] Gleick, P. H., Water management: Soft water paths. Nature, 2002. 418(6896): p. 373-373.
[26] Elderfield, H., SCHLESINGER, WH 1997. Biogeochemistry. An Analysis of Global Change, xiii+ 588 pp. San Diego, London, Boston, New York, Sydney, Tokyo, Toronto: Academic Press. Price U.S. $49.95 (paperback). ISBN 0 12 625155 X. CHAMEIDES, WL & PERDUE, EM 1997. Biogeochemical Cycles. A Computer-Interactive Study of Earth System Science and Global Change. xi+ 224 pp.+ disk. New York, Oxford: Oxford University Press. Price£ 37.50 (hardcovers). ISBN 0 19 509279 1. Geological Magazine, 1998. 135(6): p. 819-842.
[27] Broner, I., Irrigation scheduling: the water-balance approach. 1989, Colorado State University. Libraries.

[28] Farias, C. H., et al., Eficiância no uso da água na cana-de-açúcar sob diferentes lâminas de irrigação e níveis de zinco no litoral paraibano. Engenharia Agrícola, 2008. 28: p. 494-506.

[29] Häfner, K., UNESCO–United Nations Educational, Scientific and Cultural Organization, in A concise encyclopedia of the United Nations. 2010, Brill Nijhoff. p. 715-718.

[30] Loring, P. A., S. C. Gerlach, and H. P. Huntington, The new environmental security: Linking Food, water, and energy for integrative and diagnostic social-ecological research. Journal of Agriculture, Food Systems, and Community Development, 2013. 3(4): p. 55–61-55–61.

[31] John Tabak, Biofuels, Energy and the Environment, Facts on File, Infobase publishing, New York. ISBN-13: 978-0-8160-7082-4

6

Current Research Trends on the Utilization of Mono and Hybrid Nano-Fluids for Solar Energy Applications

Anil Dhanola[1] and Amit Raturi[2]

[1]Department of Mechanical Engineering, Faculty of Engineering and Technology, Guru Jambheshwar University of Science and Technology, Hisar, Haryana, India.
[2]Department of Mechanical Engineering, Govind Ballabh Pant Institute of Engineering and Technology, Pauri Garhwal, Uttarakhand India.
E-mail: anildhanola3698@gmail.com; amitraturi4493@gmail.com

Abstract

In this era, the need for energy is continuously increasing due to the fast advancements in the social economy and the growing world's population. On the other hand, the consumption of fossil fuels, and their adverse effects on the environment are a matter of concern. Solar energy is considered the best alternative to conventional energy sources and can fulfil the energy demand. The thermal performance of solar energy conversion systems is highly dependent on the working fluid. The addition of appropriate nanoparticles in base fluids has become an efficient way to enhance the thermal performances of solar systems and it has been proved that nano-fluids offer extraordinary advantages over traditional working fluids. This chapter aims to summarize the recent advancements in the applications of mono and hybrid nano-fluids in improving the efficiency of different solar energy systems. Lastly, challenging issues associated with the use of nanofluids in solar systems and future perspectives have been discussed.

Keywords: Renewable Energy, Solar Energy, Nano-fluids, Hybrid nano-fluids, Solar Energy Systems.

6.1 Introduction

Nowadays, the crisis in energy demand is increased significantly due to the continuous escalation in industries, society, and the world's population [1]. As per the International Energy Agency report, the consumption of energy in the world is likely to be raised by 44% in 2030 [2]. Also, in the coming times, traditional energy resources (such as fossil fuels) will not be able to fulfil the demand for energy due to their rapid depletion. Moreover, extensive use of these energy resources is polluting the environment and is affecting the ecosystem caused by the emission of carbon dioxide gases [3]. Keeping these issues in view, researchers and scientists have shifted their research interests towards the energy produced from renewable resources. Solar, hydropower, wind, and geothermal are common forms of renewable and sustainable energy production and also they are cheap and eco-friendly. However, among these forms, solar energy is presently the most used and is a reasonable type of renewable resource [4]. It has been observed that carbon emissions are depleted approximately by 7 lakh m^3 after installing household solar technologies in the United States of America [5]. For generating the energy of 1kWh, coal, and natural gases release carbon amounts of 1.63 kg and 0.91 kg, respectively while solar emits only 0.09 kg [6]; and for this reason, the utilization of solar energy is expanded in various applications in the last few decades such as heat and electricity generation, water heating, water desalination, building heating/cooling, etc. [7–9]. With the rapid advancements in solar energy technologies, conventional working fluids like water, air, etc. are no more efficient for heat transfer because of poor thermal conductivity and less heat-carrying capability [10, 11]. Thus, researchers are trying to find out alternate and effective methods to improve the thermal efficiency of solar systems. In recent years, nanotechnology has also been widely adopted in solar energy applications. The use of nanoparticles-based fluids (also termed nano-fluids) instead of traditional working fluids exhibit appreciable prospects to enhance the thermal performances of such systems. It has been observed that thermal conductivity and other thermal and physical attributes of base fluids can be increased by the addition of nanoparticles (size$<$100nm) [12]. Nanoparticles have impressive thermo-physical properties which help in improving the heat transfer performance of base fluids. Mostly, nanoparticles are available in spherical and cylindrical shapes and are made up of metal oxide, carbides, various forms of carbon, etc. using different chemical methodologies.

The main aim of this chapter is to present recent research and development on the applications of mono and hybrid nano-fluids for different solar energy technologies and the challenges with nano-fluids on utilizing in solar energy systems. Finally, based the literature survey, some suggestions for future work are discussed.

6.2 Nano-Fluids as Smart Fluids

Fluids containing solid metallic or non-metallic particles (having a size ranging from 1 nm to 100 nm), and are uniformly and stably dispersed in base fluids such fluids are called nano-fluids [13]. The reason behind dispersing these particles in base fluids is to improve the thermal and physical properties of base fluids. Base fluids may be in the forms of water, mineral oil, vegetable oil, refrigerant, synthetic oil, etc. Nanoparticles are added to base fluids at different concentrations. The heat transfer rate and thermal performance of the system are highly dependent on various parameters of nano-fluids such as nanoparticle shape, size, type and concentration, and base fluid properties [14].

As per the composition of nanoparticles, nanoparticles are mainly classified into three categories (Figure 6.1). Organic nanoparticles are made up of carbon-based nanomaterials for instance fullerene, single/double/multi-walled carbon nanotubes, graphite, and nanofiber. Inorganic nanoparticles include metal (Al, Zn, Fe, Ag, Cu, etc.) and oxide-based nanoparticles (Al_2O_3, TiO_2, ZnO, CuO, SiO_2, MgO, etc.). The last category is hybrid-nanoparticles which is the composite of organic and organic nanoparticles, inorganic and inorganic nanoparticles, and organic and inorganic nanoparticles.

The performance of nano-fluids is not only affected by nanoparticles physical characteristics, nanoparticles concentration, and type of base fluid but also by their preparation method. To achieve better uniform dispersion of nanoparticles in base fluids, the preparation technique plays an important role. Development of single nanoparticles-based nano-fluids or hybrid nano-fluids can be done either via single-step or double-step methods [16]. For a single-step process, the development of nanoparticles and their dispersion is carried out in a single step. This process is done either by physical or chemical techniques. Nano-fluids produced from this process showed good

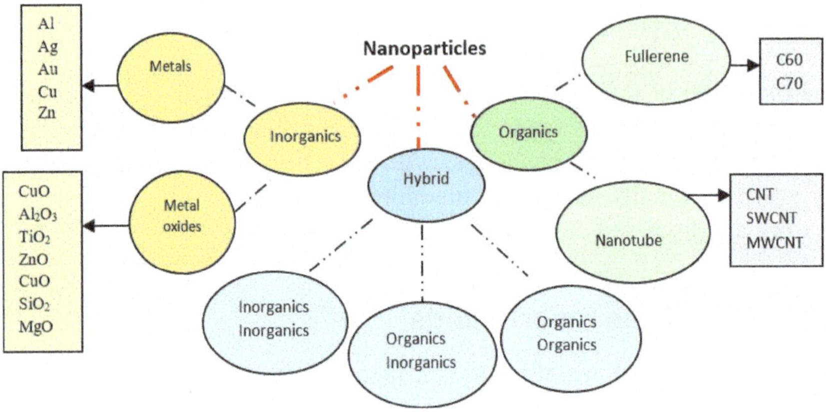

Figure 6.1 Classification of nanoparticles [15]

dispersion stability. In a two-step process, nanoparticles are produced separately by using chemical or mechanical methods, and then they are blended within the base fluids with the help of mechanical mixing or ultrasonication. Mixing time and ultrasonication time both play key roles in influencing the stability of the nanofluids. Numerous studies have shown the potential of a single component of nano-fluids in various applications including solar, tribological, cutting fluids, etc. However, due to the increased demand for versatile properties (such as rheological, thermal, tribological, etc.), single-component nano-fluids are no longer efficient. Recently, hybrid nano-fluids have gained much interest as working fluids to resolve the issue.

6.2.1 Hybrid Nano-Fluids

Nano-fluids can be categorized into two types: Mono nano-fluids and hybrid nano-fluids [17]. In mono nano-fluids, only one nanoparticle type is dispersed into the base fluid and in hybrid nano-fluids, more than one nanoparticle type is dispersed into the base fluid. It is not feasible to improve the essential properties of nano-fluids by dispersing only one nanoparticle type because some mono nano-fluids may exhibit good thermal properties but comparatively low rheological properties. Therefore, to resolve this challenge, hybrid nano-fluids are developed. With the addition of more than one nanoparticle type, a significant enhancement in thermo-physical, hydrodynamic, and heat transfer characteristics can be obtained (due to synergetic effect) in comparison to single nanoparticle type nano-fluid and traditional fluid [18]. The graphical representation of hybrid nano-fluid is represented in Figure 6.2.

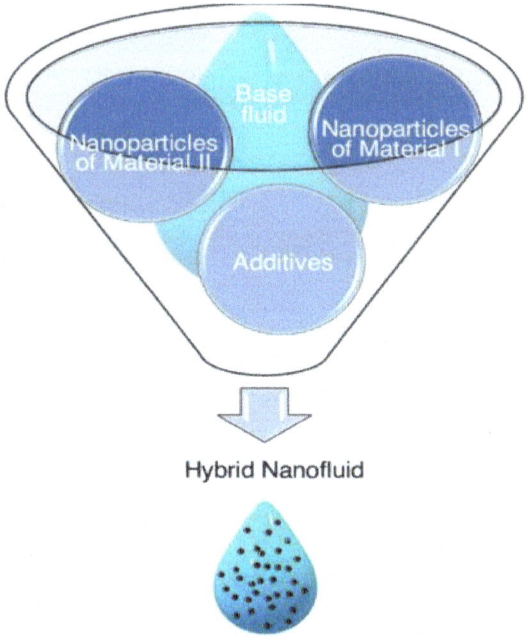

Figure 6.2 Graphical illustration of hybrid nano-fluids [19]

6.3 Utilization of Mono/Hybrid Nano-Fluids in Solar Energy Technologies

To date, a significant number of research investigations (including theoretical and experimental) have been conducted on various solar energy systems using mono and hybrid nano-fluids. This section is categorized into three subsections. In the first, second, and third subsections, recent research studies on the applications of mono/hybrid nanofluids in solar collectors, solar photovoltaic thermal (PV/T) systems, and water desalination systems are concisely discussed, respectively.

6.3.1 Solar Collectors (SCs)

Solar collectors are the kind of thermal devices that are used for heat exchange. The main function of SCs is to absorb the incoming solar radiations and turn them into heat energy. SCs consist of riser tubes in which working fluid flows and absorb solar radiation. SCs are majorly classified as non-tracking SCs and tracking SCs. Non-tracking SCs are commonly used

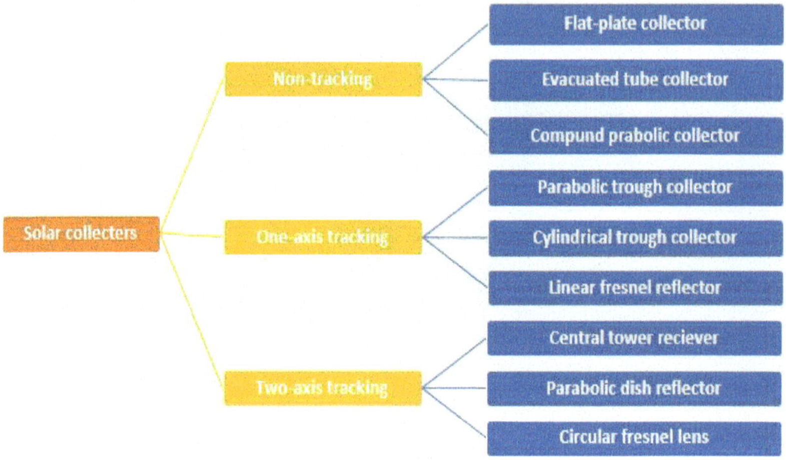

Figure 6.3 Classification of SCs [20]

SCs. Tracking SCs show better thermal efficiency and less thermal losses in comparison to Non-tracking SCs. Tracking SCs can be further classified as one-axis tracking and two-axis tracking. A detailed classification of SCs is presented in Figure 6.3. SCs are also classified based on temperature range such as low-temperature SCs (T<100 °C): Flat-plate collector; medium temperature SCs (Approx. 400 °C): Parabolic trough and linear Fresnel SCs; high-temperature SCs (T>400 °C): Dish and power tower SCs [21, 22].

Often water is used as a working fluid in SCs. However, water exhibits poor thermal performance due to limited absorption range. Therefore, researchers are putting their efforts to develop new smart fluids. Based on earlier published research, researchers have suggested that the thermal performances of working fluids can be improved by dispersing nanoparticles, and it is an efficient method to improve SCs efficiency [23].

The key findings of several recent studies on the performance enhancement of various solar collectors using mono and hybrid nano-fluids are summarized in Table 6.1. Overall outcomes from these studies stated that the addition of nanoparticles is found favorable in improving the efficiency of solar collectors when compared with conventional working fluids. In addition, many researchers have also compared the performance of mono and hybrid nano-fluids and observed that hybrid nano-fluids are more efficient w.r.t. mono nano-fluids and conventional fluids in solar-thermal conversion. As a result, researchers are taking more research interest in hybrid nano-fluids in recent years for solar energy applications.

Table 6.1 Summary of recent studies on solar collectors using mono/hybrid nano-fluids

Ref. No.	Nano-fluid details	Type of solar collector	Key findings
24	Hybrid (TiO_2-Ag/Water)	Volumetric	Temperature of hybrid nano-fluid was found much greater than that of TiO_2/Water mono nano-fluid. However, the temperature of Ag/Water mono nano-fluid was found equal to the hybrid nano-fluid.
25	Mono (SiO_2/Ethylene glycol-Water)	Flat-plate	Efficiency of the system increased with the rising of nanoparticles concentrations ranging from (0-1%) and enhancement was found from 4% to 8%. Variation in efficiency was negligible between 0.75% and 1% concentration.
26	Mono (Graphene/Ionic liquid)	Direct absorption	Results showed that experimental temperature agreed with simulated results. Efficiency of the solar system increased with solar concentration and receiver height but decreased with the addition of graphene nanomaterials.
27	Mono (MWCNT, Al_2O_3, CuO, SiO_2 TiO_2/Propylene glycol-Water)	U-tube	Results stated that the thermal conductivity of different nano-fluid samples increased with increasing nanoparticle concentration. Moreover, thermal efficiency (from high to low) of solar collector followed trend as: nano-fluid$_{MWCNT}$> nano-fluid$_{CuO}$> nano-fluid$_{Al_2O_3}$>nano-fluid$_{TiO_2}$>nano-fluid$_{SiO_2}$.
28	Mono (Silver/Water)	Direct absorption	The efficiency of the collector increased with increasing collector height and concentration of silver nanomaterials. However, an optimum height of 10 mm and concentration of 0.03% were recorded optimum at which maximum efficiency (90%) was achieved.
29	Mono (Plasmonic nano-fluid)	Flat-plate	Simulated results showed that the efficiency of the solar collector increased with mass flow rate while it decreased with increasing channel length.

(Continued)

Table 6.1 Continued.

Ref. No.	Nano-fluid details	Type of solar collector	Key findings
30	Mono (SiO_2/Water)	Evacuated tube	Heat transfer characteristic of nano-fluid was observed higher than that of water. After 3 hrs of solar radiation, the temperature of nano-fluid was received higher as compared to water.
31	Mono (CeO_2/Water)	Flat-plate	Both numerical and experimental results revealed that water-based nano-fluid containing CeO_2 nanoparticles with 0.01% concentration improved the efficiency of the collector by 21.5% w.r.t. water.
32	Mono (Al_2O_3/Water)	Flat-plate and U-tube	An increment in performance of flat-plate and U-tube solar collectors was observed at approx..15% and 11%, respectively. The concentration of 1 vol. % of Al_2O_3 nanoparticles was obtained optimum at which maximum efficiency was observed for the collectors.
33	Mono (Al_2O_3, TiO_2/Water) Hybrid (Al_2O_3-TiO_2/Water)	Flat-plate	Results revealed that thermal performance of flat-plate solar collector was improved by 19%, 21%, and 26% while using Al_2O_3/Water nano-fluid, TiO_2/Water nano-fluid, and Al_2O_3-TiO_2/Water hybrid nano-fluid, respectively. On comparing data, less deviation was observed between simulated results and experimental results. The authors suggested using hybrid nano-fluid from an economical and efficient point of view.
34	Hybrid (ZnO-Au/Oil)	Direct absorption	Based on the results, the authors highly recommended the prepared hybrid nanofluid as working fluid to the direct absorption solar collector as 240% enhancement in photo-thermal conversion efficiency was observed w.r.t. base fluid.
35	Mono (Barely husk-TiO_2 and Olive leaves extract- TiO_2/Water	Parabolic trough	Both samples of nano-fluid showed better thermal and energetic performance in comparison to the base fluid. Increased thermal efficiency of

Table 6.1 Continued.

Ref. No.	Nano-fluid details	Type of solar collector	Key findings
			0.073% and 0.077% was observed for barely husk-TiO_2/Water and Olive leaves extract-TiO_2/Water, respectively.
36	Hybrid (MWCNT-SiO_2/Ag/Water)	Volumetric	Prepared samples of hybrid nano-fluids (HNFs) exhibited better solar-thermal conversion as compared to mono nano-fluids. This result was due to the higher thermal conductivity of HNFs. The authors recommended HNFs for volumetric SCs.
37	Mono (Al_2O_3, TiO_2/ Oil) Hybrid (TiO_2-Al_2O_3/ Oil)	Parabolic trough	Hybrid nano-fluid was found more efficient than both mono nano-fluids in terms of thermal performance. Mono nano-fluids showed similar thermal efficiency (i.e. 0.34%) while 0.74% was observed for hybrid nano-fluid.
38	Hybrid (Fe_3O_4-SiO_2/Water	Direct absorption	The experimental study's results revealed that the efficiency of SiO_2/Water nano-fluid found 11% and 28% lower than that of water and hybrid nano-fluid, respectively.
39	Mono (Al_2O_3/ Water)	Flat-plate	Prepared nano-fluid exhibited better thermal efficiency at flow rates below 0.016 kg/s. The highest efficiency of 74% (at 3 vol. %) and 84% (at 1 vol. %) was observed in July and October respectively at a fixed flow rate.
40	Hybrid (MgO-MWCNT, CuO-MWCNT/Water	Flat-plate	Experimental results showed that the solar collector's thermal efficiency was increased by 18% and 20.52% by using CuO-MWCNT/Water HNF and MgO-MWCNT/ Water HNF, respectively. Exegetic performance was also obtained higher for MgO-MWCNT/ Water HNF.
41	Mono (CuO/ Water)	Flat-plate	The authors experimentally investigated the thermal performance of solar collectors using CuO/ Water nano-fluid at different flow rates (1-4 L/min).

(Continued)

Table 6.1 Continued.

Ref. No.	Nano-fluid details	Type of solar collector	Key findings
			Collector efficiency was obtained higher using nano-fluid w.r.t. water. Also, efficiency was increased with the rise of flow rate and found to a maximum of 55% at 4 L/min.
42	Mono (CuO/ Water)	Evacuated tube	Authors experimentally studied the thermal efficiency of a collector using different sizes of CuO nanoparticles/ Water nano-fluid and water. Results showed that the efficiency was improved by incorporating nanoparticles within the water. Also, nano-fluid containing lower sized nanoparticles (40 nm) exhibited better thermal performance in comparison to higher sized nanoparticles (80 nm).
43	Mono (MWCNT/Water)	Flat-plate	The thermal efficiency of the selected solar collector was improved with the addition of MWCNT and it was increased with the increase of nanoparticles concentration. At the highest concentration (i.e., 0.1 wt. %), efficiency raised by approx. 34% w.r.t. water. As per the experimental study, authors found MWCNT/Water nano-fluid suitable for improving energy and exergy performance of flat-plate SC
44	Hybrid (TiO_2-Al_2O_3/ Therminol-55)	Parabolic trough	The synergetic effect of both nanoparticle types play important role in increasing thermal conductivity (TC) and thus TC was increased by 35% for hybrid nano-fluid. At a concentration of 0.5 wt. %, hybrid nano-fluid showed maximum TC.
45	Mono (Al_2O_3/Water) and Hybrid (Al_2O_3-Fe/ Water)	Flat-plate	A numerical study revealed that mono nano-fluid exhibited better thermal performance in the collector in comparison to hybrid nano-fluid, as thermal efficiency increased by 2.16% w.r.t. water.

Table 6.1 Continued.

Ref. No.	Nano-fluid details	Type of solar collector	Key findings
46	Mono (Al_2O_3, CeO_2, CuO/ Syltherm 800) Hybrid (Al_2O_3-$CeO_2$and Al_2O_3- CuO/ Syltherm 800)	Parabolic trough	According to a theoretical study, a favorable increment in thermal efficiency in solar collectors using both mono and hybrid nano-fluids was observed. However, maximum enhancement in thermal efficiency was obtained for hybrid nano-fluids. For Al_2O_3-CeO_2 and Al_2O_3-CuO/Syltherm 800, it was increased by 1.08% and 1.07%, respectively.
47	Mono (MWCNT, TiO_2, SiO_2 and Cu/ Water)	Evacuated tube	The theoretical study revealed that the heat transfer rate is increased with the rise in nanoparticle concentration ratios (ranging from 0.5 5%) for all mono nano-fluids types. Cu/Water nano-fluid showed better thermal performance in the solar collector, and the highest heat transfer of 14.09% was observed for this sample at a 5% concentration of Cu.
48	Hybrid (Al_2O_3-Fly ash and SiO_2-Fly ash/ water)	Direct absorption	It was observed that a hybrid composite of Al_2O_3 /fly ash showed better thermal efficiency (i.e. 72.82%) while SiO_2/ fly ash reveals 59.23%. Also, thermo-physical attributes of nano-fluids increased with alumina concentration while decreased with silica concentration.
49	Hybrid (Ag-ZnO, Ag-TiO_2, Ag-MgO/ Syltherm 800)	Parabolic trough	As per the numerical study, the heat transfer for sample Ag-ZnO, Ag-TiO_2, and Ag-MgO/Syltherm 800 were increased by 26%, 29%, and 31% respectively w.r.t. neat base fluid at volume fraction (%) of 4 and hence thermal efficiency was also obtained for this sample. An increment in friction factor was also observed higher for hybrid nano-fluids.

(Continued)

Table 6.1 Continued.

Ref. No.	Nano-fluid details	Type of solar collector	Key findings
50	Hybrid (Nano diamond-Cobalt oxide/Water)	Flat-plate	The experimental study revealed that the combination of ND-Co_3O_4 increased viscosity and thermal conductivity of nanofluids, and viscosity and thermal conductivity increased by approximately 16% and 46% respectively at 0.15 wt. % concentration and temperature 60 °C. The thermal efficiency of the collector was increased by 20% using a hybrid nano-fluid at 0.15 wt. % concentration of ND-Co_3O_4.
51	Hybrid (rGo-Co_3O_4/Water)	Linear Fresnel	Thermal efficiency of linear Fresnel was increased from approx. 3% to 32% during using hybrid nano-fluids at 0.2 wt. % of rGo-Co_3O_4. A significant improvement in viscosity, thermal conductivity, and density was also obtained after dispersing a hybrid composite of rGo-Co_3O_4.
52	Mono (Cu, SiO_2/ Water) Hybrid (Cu- SiO_2/ Water)	U-tube	Thermal performance of solar collector was improved by 15% using nano-fluids w.r.t. water only. The authors suggested using hybrid nano-fluid to overcome the problem of precipitation and to improve the thermal performance of the system.

6.3.2 Photovoltaic Thermal (PV/T) System

The use of photovoltaic (PV) modules is the best way to generate electrical power through solar energy. The key challenge with the PV module is that, with an increase in the intensity of solar light, there is an increased temperature of the module and due to this result, a significant decrement in the power generation efficiency of the PV system takes place. Therefore, to overcome this issue, photovoltaic thermal (PV/T) systems are proposed. In these systems, a working fluid (water/oil/air) is used that not only reduced the module's temperature but is also used for thermal applications [53]. In addition, reducing the module's temperature leads to generating high electrical power. Some schematic diagrams of different PV/T systems are depicted in Figure 6.4. In previous studies, it has been proved that the efficiency of PV/T

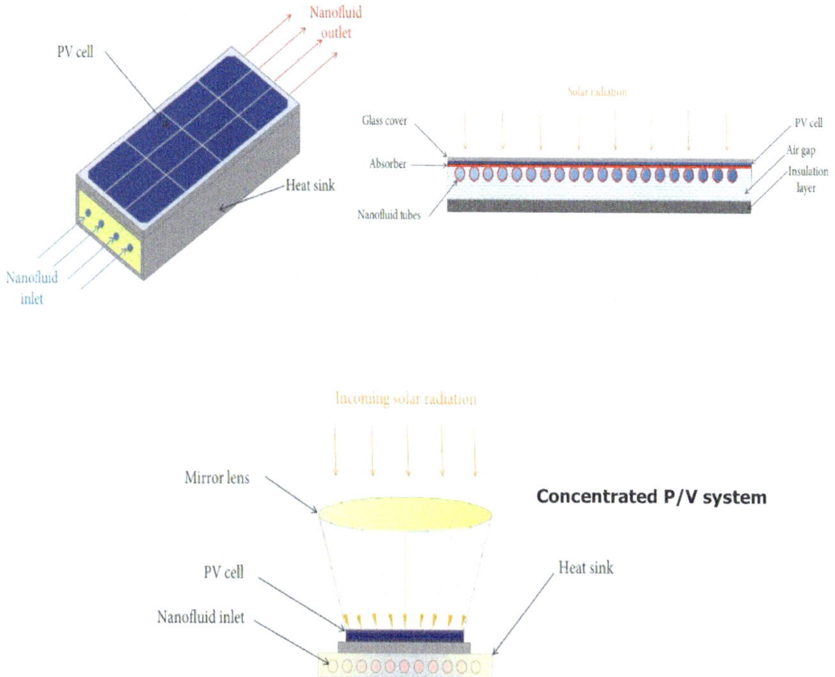

Figure 6.4 Schematic diagrams of PV/T systems [55]

systems and the heat transfer process can be further enhanced by using nano-fluids. This is only because of the improved thermal and physical properties of nano-fluids w.r.t. base fluid [54].

Table 6.2 presents the summary of recent studies on PV/T systems using mono and hybrid nano-fluids. It is observed that the photothermal conversion efficiency of PV/T has greatly improved using nano-fluids in comparison to traditional working fluids. Researchers have done sufficient research works on mono nano-fluids however very limited work is available on hybrid nano-fluids. So, researchers need to pay more attention to hybrid nano-fluids for PV/T systems, as they are more efficient than mono nano-fluids in improving the efficiency of the system.

6.3.3 Solar Desalination

Due to the rapid industrial development, the concern of global warming and the increase in the world's population, the demand for freshwater is increasing day by day. As a result of this, water resources are being depleted and the

Table 6.2 Summary of recent studies on PV/T systems using mono/hybrid nano-fluids

Ref. No.	Nano-fluid details	Key findings
56	Au, Ag, and Au-Ag based nano-fluids	The authors observed from the investigation that the photothermal conversion efficiency and temperature improved with increasing Au nanoparticles. The adopted preparation method was considered the best method for preparing nano-fluids as working fluids for PV/T systems.
57	Mono (Gold/water)	Using different sizes of Au nanoparticles in nano-fluids revealed no effect on affecting photo-thermal conversion efficiency. Moreover, higher efficiency was observed at lower concentrations of Au nanoparticles.
58	Mono (Ag/Water, Al_2O_3/Water)	Numerical results showed that efficiency and heat transfer coefficient (HTC) of PV/T system increased with increasing nanoparticles concentration for both types of nano-fluid. However, maximum HTC was found approx. 43% with Ag /Water nano-fluid.
59	Mono (Al_2O_3, CuO, SiC/Water)	Experimental results showed that all the prepared nano-fluids of different types of nanoparticles had better thermal conductivity in comparison to the water. However, among these nano-fluids, nano-fluid containing SiC nanoparticles showed better thermal conductivity in PV/T system w.r.t. other nano-fluids due to higher stability.
60	Hybrid (CuO-ZnO/Water)	Photo-thermal conversion outcomes showed that the temperature and thermal conductivity of hybrid nano-fluid was found better than that of CuO/Water nano-fluid and the maximum temperature was reached 73% at 0.01 vol. %.
61	Hybrid (Fe_3O_4-SiO_2/Water)	The photo-thermal conversion efficiency of water was greatly improved by the addition of Fe_3O_4-SiO_2 nanoparticles and this was because of silica coating, which improved the temperature of nano-fluid.
62	Mono (TiO_2/Polyethylene glycol-Water, Al_2O_3/ Cetyltrimethylammonium bromide-Water)	Both nano-fluids successfully decreased the PV cell's temperature and were found better than that water. However, Al_2O_3 mono nano-fluid provided better performance than

Table 6.2 Continued.

Ref. No.	Nano-fluid details	Key findings
		other types of nano-fluid. Moreover, samples containing a higher percentage of nanoparticles exhibited better Colling of PV cells.
63	Hybrid (Al_2O_3-ZnO/Water)	The photo-thermal conversion efficiency of the PV/T system was observed higher in using hybrid nano-fluid in comparison to water. The efficiency and exergy were increased by 4.1% and 4.6% at 0.05 wt. % of nanoparticles.
64	Mono (MWCNT, TiO_2/ Water)	Both nano-fluids decreased more temperature of the PV module in comparison to the water only. At 0.1 wt. %, TiO_2-based nano-fluid showed the lowest temperature (i.e., 2.01 °C) of PV module
65	Hybrid (CNT-Al_2O_3/Water)	It was observed that the electrical efficiency improved by 7.15% to 8.2% on using water and hybrid nano-fluid as a working fluid, and the overall efficiency of PV/T increased by 27% w.r.t water.
66	Hybrid (ZnO-Al_2O_3/Water)	Experimental results showed that maximum efficiency of the PV/T system was recorded when a mixture ratio of 0.47 of Al_2O_{3was}added to the hybrid nano-fluid. Electrical, thermal and exergy efficiencies of the system increased by approx. 14%, 56%, and 15%, respectively. Further, the temperature of the PV module was decreased with the increasing mass flow rate of hybrid nano-fluids.
67	Hybrid (CuO-MgO-TiO_2/Water)	Numerical results revealed that the maximum efficiency of the PV/T system was obtained at a concentration of 0.01 vol. % and electrical, thermal and exergy efficiencies increased by approx. 14%, 58%, and 16% respectively on using composites of CuO-MgO-TiO_2 in the water. Pressure drop and pumping power were also observed lower by using composites.

crisis of drinking water is at its peak. About 70% of the earth's surface is water-covered, most of this is not potable. Moreover, contaminated water in form of rainwater, boring wells, rivers, lake, and seawater may cause

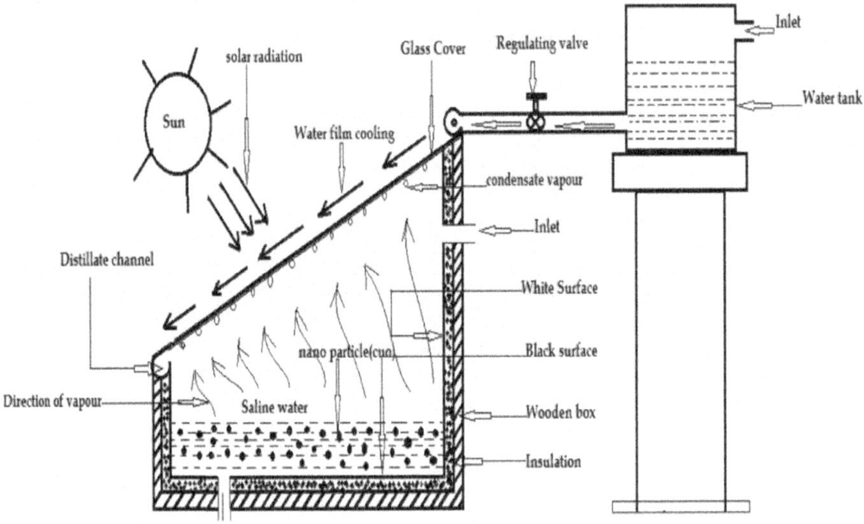

Figure 6.5 Schematic view of solar still using nano-fluids [71]

serious health-related issues on consumption. Thus, contaminated water must be impurity-free [68]. Solar still is one of the affordable, effective, natural, and eco-friendly water desalination technologies to get potable water [69]. However, the use of solar still is limited due to the low productivity rate. This low productivity is the result of loss of heat to the surroundings and requires a plenty portion of the heat to heat the basin water [70]. Solar still with using nano-fluids is shown in Figure 6.5.

As per the literature, the use of nano-fluids has been found one of the efficient approaches to overcoming these issues. Nano-fluids have superior thermophysical and optical characteristics which help in increasing the heat transfer process and reducing heat loss inside the solar still basin. Due to this fact, the productivity of solar stills is increased. In recent years, research on the performance of solar still using mono nano-fluids and hybrid nano-fluids has become a key interest of researchers. Summary of some recent investigations is tabulated in Table 6.3.

6.4 Challenges with Nano-Fluid-Based Solar Technologies

In the last few years, substantial research works related to the utilization of mono/hybrid nano-fluids in solar energy technologies have been carried out, and have established advantageous solutions for improving thermal,

Table 6.3 Summary of recent studies on solar stills using mono/hybrid nano-fluids

Ref. No.	Nano-fluid details	Key findings
72	Mono (ZnO, Al_2O_3, Fe_2O_3, SnO_2/ Water)	Results showed that nano-fluids having good thermal conductivity absorb more heat from the sun. On using Al_2O_3/Water nano-fluid, still, productivity increased by approx. 30% while with ZnO and SnO_2 nano-fluids, production was approx. 13% and 19% are respectively w.r.t. water.
68	Mono (Al_2O_3,/ Water)	Increment in productivity increased with rising nanoparticles concentration. At the highest concentration (0.12%), the increment was observed approx. 12% and 8% for 25 kg and 80 kg of water respectively w.r.t. water.
73	Mono (Al_2O_3, CuO, TiO_2/ Water)	Outcomes revealed that, at an optimized concentration of 0.25% of nanoparticles, Al_2O_3 nano-fluid gave more thermal energy efficiency (i.e. 50.34%) followed by 46.10% for TiO_2 nano-fluid and 43.81% for CuO nano-fluid w.r.t. water.
74	Mono (SiO_2, Cu/Water)	At lower temperatures, the performance of solar still in terms of evaporation rate was found to maximum using Cu/Water nano-fluid w.r.t. SiO_2/Water but at higher temperatures, this trend was observed the opposite.
75	Mono (Copper oxide, Graphite micro-flakes/ Water)	Productivity of solar still without using glass cover cooling increased by 44.91% and 53.95% with copper oxide nano-fluid and Graphite nano-fluid respectively in comparison to water. Productivity of modified solar still (including glass cover cooling) was obtained at 47.80% and 57.60% for copper oxide nano-fluid and Graphite nano-fluid respectively.
76	Mono (TiO_2, CuO, GO/ Paraffin)	Productivity of solar still was obtained as: 3.92, 4.94, 5.28, and 3.66 lit/m^2/day for paraffin, TiO_2 nano-fluid, CuO nano-fluid, and GO nano-fluid, respectively. As per economic analysis, the lowest cost was observed for CuO nano-fluid (i.e., 0.026$). As per cost analysis and productivity analysis, the authors recommended CuO nano-fluid for better performance of solar still.
77	Mono (Magnetic MWCNTs/ Saline water)	Results showed that, at 0.04 wt. % magnetic MWCNTs nanoparticles, nano-fluid absorbs almost 100% heat from the sun when fluid's thickness increased 1 cm. Evaporation efficiency increased from 24.91% to 76.65%, when concentration varied from 0 wt. % to 0.04 wt.%.
78	Hybrid (Al_2O_3-SiO_2, Al_2O_3-CuO/ Water)	Authors suggested hybrid nano-fluid for solar still in the place of mono nano-fluids for better performance. On using Al_2O_3-SiO_2 hybrid nano-fluid, maximum pure water productivity of approx. 5 kg/ m^2/day was obtained. For the same hybrid nano-fluid, thermal and exergy increased by 37.76% and 0.82%, respectively.

economic, and environmental performances of various solar energy systems. However, there are still a few issues and limitations with nano-fluids that may limit the practical application of nanofluids in solar energy systems. These challenges mainly include [18], [79]: (a) Instability, (b) Increased pressure drop and pumping power, (c) Higher production costs, (d) Inaccuracies in proposed thermo-physical models. Poor dispersion stability of nano-fluids is one of the major challenges so far. Nanoparticles tend to stick with each other due to strong Van der Waals interactions, which leads to the formation of large clusters, and then they tend to settle down. To overcome this instability issue, physical methods (magnetic stirring, ultrasonication, etc.) and chemical methods (use of surfactants and surface treatment of nanoparticles) have been proposed. This similar challenge is also faced by hybrid nano-fluids, and the rate of settling down of nanoparticles may be more for hybrid nano-fluids w.r.t. mono nano-fluids. This is due to the distinct properties of nanoparticles, which varies from one type of nanoparticles to another one. Moreover, the optimum procedure for selecting suitable nano-materials for the development of hybrid nano-fluids is still a matter of concern for researchers. Therefore, despite the impressive properties of mono nano-fluids, hybrid nano-fluids have not gained much popularity. Increased viscosity of nano-fluids due to the addition of nanoparticles may increase pumping power, pressure drop, and friction factor. During the application, nano-fluids require more energy to flow as compared to conventional fluids due to the pressure drop. In addition, friction factor is also increased due to pressure drop as a result erosion of solar system's pipes may occur. The production process of hybrid nano-fluids is generally more complex in comparison to the production of mono nano-fluids. That's why the production cost of hybrid nano-fluids is more. Most of the proposed correlations in existing studies for predicting thermo-physical properties fail to predict experimental results precisely. Therefore, these flaws will lead to inaccurateness in numerical modeling of the performance of nano-fluid-based solar systems. Discharging nano-fluids into the environment during the cleaning process is another issue that may harmful to our ecosystem.

6.5 Conclusions and Future Outlook

In the current study, an up-to-date review of recent studies related to applications of mono/hybrid nano-fluids in solar energy technologies is carried out. As per available literature, it has been observed that the applications of mono nano-fluids and hybrid nanofluids as working fluids significantly improve the overall performance of solar energy systems. Based on the review of recent studies, the following conclusions and future works can be revealed.

- Based on experimental and theoretical studies, it has been proved that hybrid nano-fluids are more effective and efficient (due to synergetic effects) than mono nano-fluids in enhancing the thermal efficiency of solar systems. However, the scope of hybrid nano-fluids for commercial purposes is restricted due to various challenges such as poor stability, increased pressure drop and friction factor, cost of production process etc. These challenges need to be considered in future studies.
- Plenty of research studies have been carried out on different solar collectors using both mono nano-fluids and hybrid nano-fluids. However, very limited studies have been investigated the performances of PV/T systems and solar desalination systems using hybrid nano-fluids. Therefore, it is suggested that researchers should explore the capability of hybrid nano-fluids in such systems also.
- In many studies, it is observed that beyond a certain concentration of nanoparticles no further enhancement in the performance of solar systems is observed. Therefore, the optimum concentration of nanoparticles must be tested. Further, the size of nanoparticles is another important parameter through which the efficiency of solar systems may affect. Thus, this parameter must be considered in further investigations.
- The higher production cost of nano-fluids is another challenging issue. Therefore, the cost-based analysis should be considered in the research.
- It is noticed that, in published research works, the performance of solar energy systems by using oxide-based nano-fluids has been widely studied. Therefore, in the coming times, more experimental and theoretical studies need to be conducted with different kinds of nanomaterials aiming to improve solar energy systems efficiency.

References

[1] M. H. Rezaei, M. Sadeghzadeh, M. AlhuyiNazari, M.H. Ahmadi and F.R.Astaraei, "Applying GMDH artificial neural network in modelling CO_2 emission in four Nordic countries," International Journal of Low-Carbon Technologies, vol. 13, pp. 266-271, 2018.

[2] Energy Information Administration's (EIA), Center for strategic and international studies, October 14, 2020, Washington, DC. Online. Available: https://www.eia.gov/outlooks/ieo/index.php.

[3] M. H. Ahmadi, M. AlhuyiNazari, R. Ghasempour, F. Pourfayaz and T. Ming, "A review on solar-assisted gas turbines," Energy Science and Engineering, vol. 6, pp. 658-674, 2018.

[4] N. Kannan and D. Vakeesan, "Solar energy for future world: A review," Renewable and Sustainable Energy Reviews, Vol. 62, pp. 1092-1105, 2016.

[5] Mashail S. Arif, "Residential solar panels and their impact on the reduction of carbon emissions," University of California, Berkeley, pp. 1-18, 2013.

[6] National Renewable Energy Laboratory: Solar has the most Potential of any renewable energy source, Think progress, July 30, 2012. Online. Available: https://archive.thinkprogress.org/national-renewable-energy-laboratory-solar-has-the-mostpotential-of-any-renewable-energy-source-87da2c774fcc/.

[7] S. Mekhilef, R. Saidur and A. Safari, "A review on solar energy use in industries," Renewable and Sustainable Energy Reviews, Elsevier, Vol. 15(4), pp. 1777-1790, 2011.

[8] S. Hoseinzadeh and R. Azadi, "Simulation and optimization of a solar-assisted heating and cooling system for a house in Northern of Iran," Journal of Renewable and Sustainable Energy, Vol. 9, 045101, 2017.

[9] O. Bamisile, A. Babatunde, M. Dagbasi and I. Wole-Osho, "Assessment of solar water heating in Cyprus: Utility, development and policy," International Journal of Renewable Energy Research, Vol. 7(3), pp.1448-1453, 2017.

[10] Y. Tripanagnostopoulos, "Aspects and improvements of hybrid photovoltaic/thermal solar energy systems," Solar Energy, Vol. 81(9), pp. 1117-1131, 2007.

[11] S. R. Reddy, M. A. Ebadian and C. X. Lin, "A review of PV–T systems: thermal management and efficiency with single phase cooling," International Journal of Heat and Mass Transfer, Vol. 91, pp. 861–871, 2015.

[12] S. U. S. Choi and J. A. Eastman, "Enhancing thermal conductivity of fluids with nanoparticles," ASME International Mechanical Engineering Congress & Exposition, pp. 1-9, 1995.

[13] R. Hajian, M. Layeghi and K. Abbaspour Sani, "Experimental study of nanofluid effects on the thermal performance with response time of heat pipe," Energy Conversion and Management, Vol. 56, pp. 63–68, 2012.

[14] N. Chavda, "Review of modeling the performance of heat pipes using artificial neural network," International Journal of Artificial Intelligence and Knowledge Discovery, Vol.7 (3), pp. 7-12, 2017.

[15] H. H. Al-Kayiem, S. C. Lin and A. lukmon, "Review on nanomaterials for thermal energy storage technologies," Nanoscience & Nanotechnology-Asia, Vol. 3(1), pp. 60-71, 2013.

[16] W. Yu and H. Xie, "A review on nanofluid: preparation, stability mechanisms, and applications," Journal of Nanomaterials, Vol. 2012, pp. 1-17, 2012.

[17] M. H. Ahmadi, A. Mirlohi, M. Alhuyi Nazari and R. Ghasempour, "A review of thermal conductivity of various nano fluids," Journal of Molecular Liquids, Vol. 265, pp. 181-188, 2018.

[18] J. Sarkar, P. Ghosh and A. Adil, "A review on hybrid nanofluids: recent research, development and applications," Renewable and Sustainable Energy Reviews, Vol. 43, pp. 164-177, 2015.

[19] H. Babar, M. Usman Sajid and H. M. Ali, "Viscosity of hybrid nanofluids: A critical review," Thermal Science International Scientific Journal, Vol. 23 (3), pp. 1713-1754, 2019.

[20] E. Cuce, P. M. Cuce, T. Guclu and A. B. Besir, "On the use of nanofluids in solar energy applications," Journal of Thermal Science, Vol. 29(4), pp. 513-534 , 2020.

[21] D. Vallentin and P. Viebahn, "Economic opportunities resulting from a global deployment of concentrated solar power (CSP) technologies—the example of German technology providers," Energy Policy, Vol.38 (8), pp. 4467-4478, 2010.

[22] K. S. Reddy, N. R. Kamnapure and S. Srivastava, "Nanofluid and nanocomposite applications in solar energy conversion systems for performance enhancement: a review," International Journal of Low-Carbon Technologies, Vol. 12 (1), pp. 1-23, 2017.

[23] A. H. Elsheikh, S. W. Sharshir, M. M. Mostafa, F. A. Essa, A. Ali and M. Kamal, "Applications of nanofluids in solar energy: A review of recent advances," Renewable and Sustainable Energy Reviews, Vol. 82, pp. 3483-3502, 2018.

[24] Y. Xuan, H. Duan and Q. Li, "Enhancement of solar energy absorption using plasmonic nanofluid based on TiO_2/Ag composite nanoparticles," Royal Society of Chemistry Advances, Vol. 4, pp. 16206-16213, 2014.

[25] S. S. Meibodi, A. Kianifar, H. Niazmand, O. Mahian and S. Wongwises, "Experimental investigation on the thermal efficiency and performance characteristics of a flat plate solar collector using SiO_2/EG–water nanofluids," International Communications in Heat and Mass Transfer, Vol. 65, pp. 71-75, 2015.

[26] J. Liu. Z. Ye, L. Zhang, X. Fang and Z. Zhang, "A combined numerical and experimental study on graphene/ ionic liquid nano fluid based direct absorption solar collector," Solar Energy Materials and Solar Cells, Vol. 136, pp. 177-186, 2015.

[27] H. Kim, J. Ham, C. Park and H. Cho, "Theoretical investigation of the efficiency of a U-tube solar collector using various nanofluids," Energy, Vol. 94, pp.497-507, 2016.

[28] M. Chen, Y. He, J. Zhu and D. Wen, "Investigating the collector efficiency of silver nanofluids based direct absorption solar collectors," Applied Energy, Vol. 181, pp. 65-74, 2016.

[29] J. Jeon, S. Park and B. J. Lee, "Analysis on the performance of a flat-plate volumetric solar collector using blended plasmonic nanofluid," Solar Energy, Vol. 132, pp. 247-256, 2016.

[30] S. Yan, F. Wang, Z. Shi and R. Tian, "Heat transfer property of SiO_2/water nanofluid flow inside solar collector vacuum tubes," Applied Thermal Engineering, Vol. 118, pp. 385-391, 2017.

[31] P. Michael Joseph Stalin, T. V. Arjunan, M. M. Matheswaran and N. Sadanandam, "Experimental and theoretical investigation on the effects of lower concentration CeO_2 / water nano fluid in flat plate solar collector," Journal of Thermal Analysis and Calorimetry, Vol.135, pp. 29-44, 2019.

[32] W. Kang, Y. Shin and H. Cho, "Economic Analysis of Flat-Plate and U-Tube Solar Collectors Using an Al_2O_3 Nanofluid," Energies, Vol. 10, pp. 1-15, 2017.

[33] E. Farajzadeh, S. Movahed and R. Hosseini, "Experimental and numerical investigations on the effect of Al_2O_3/ TiO_2- H_2O nano fluids on thermal efficiency of the flat plate solar collector," Renewable Energy, Vol. 118, pp. 122-130, 2018.

[34] X. Wang, Y. He, M. Chen and Y. Hu, "ZnO- Au composite hierarchical particles dispersed oil-based nanofluids for direct absorption solar collectors," Solar Energy Materials and Solar Cells, Vol. 179, pp. 185-193, 2018.

[35] E. C. Okonkwo, E. A. Essien, E. Akhayere, M. Abid, D. Kavaz and Tahir A.H. Ratlamwala, "Thermal performance analysis of a parabolic trough

collector using water-based green-synthesized nanofluid," Solar Energy, Vol. 170, pp. 658-670, 2018.

[36] J. Zeng and Y. Xuan, "Enhanced solar thermal conversion and thermal conduction of MWCNT-SiO_2/Ag binary nanofluids," Applied Energy, Vol. 212, pp. 809-819, 2018.

[37] E. Bellos and C. Tzivanidis, "Thermal analysis of parabolic trough collector operating with mono and hybrid nanofluids," Sustainable Energy Technologies and Assessments, Vol. 26, pp. 105-115, 2018.

[38] M. Karami, "Experimental investigation of first and second laws in a direct absorption solar collector using hybrid Fe_3O_4/SiO_2 nanofluid," Journal of Thermal Analysis and Calorimetry, Vol. 136, pp. 661–671, 2019.

[39] A. M. Genc, M. A. Ezan and A. Turgut, "Thermal Performance of a nano fluid based flat plate solar collector: A transient numerical study," Applied Thermal Engineering, Vol. 130, pp. 395-407, 2018.

[40] S. K. Verma, A. K. Tiwari, S. Tiwari and D. S. Chauhan, "Performance analysis of hybrid nanofluids in flat plate solar collector as an advanced working fluid," Solar Energy, Vol. 167, pp. 231-241, 2018.

[41] M. Mirzaei, "Experimental investigation of CuO nanofluid in the thermal characteristics of a flat plate solar collector," Environmental Progress & Sustainable Energy, Vol. 38, pp. 260-267, 2019.

[42] W. Kang, Y. Shin and H. Cho, "Experimental investigation on the heat transfer performance of evacuated tube solar collector using CuO nanofluid and water," Journal of Mechanical Science and Technology, Vol. 33, pp. 1477–1485, 2019.

[43] M. Eltaweel and A. A. Abdel-Rehim, "Energy and exergy analysis of a thermo siphon and forced-circulation flat-plate solar collector using MWCNT/Water nanofluid," Case Studies in Thermal Engineering, Vol. 14, pp. 1-11, 2019.

[44] O. Gulzar, A. Qayoum and R. Gupta, "Experimental study on thermal conductivity of mono and hybrid Al_2O_3- TiO_2 nano fluids for concentrating solar collectors," International Journal of Energy Research, Vol. 45, pp. 4370-4384, 2021.

[45] E. C.Okonkwo, I. Wole-Osho, D. Kavaz, M. Abid and T. Al-Ansari, "Thermodynamic evaluation and optimization of a flat plate collector operating with alumina and iron mono and hybrid nanofluids," Sustainable Energy Technologies and Assessments, Vol. 37, pp.1-14, 2020.

[46] O. Al- Oran, F. Lezsovits and A. Aljawabrah, "Exergy and energy amelioration for parabolic trough collector using mono and hybrid nanofluids," Journal of Thermal Analysis and Calorimetry, Vol. 140, pp.1579–1596, 2020.

[47] A. Yurddas, "Optimization and thermal performance of evacuated tube solar collector with various nanofluids," International Journal of Heat and Mass Transfer, Vol. 152, 119496, 2020.

[48] P. P. Thakur, T. S. Khapane and S. S. Sonawane, "Comparative performance evaluation of fly ash-based hybrid nanofluids in microchannel-based direct absorption solar collector," Journal of Thermal Analysis and Calorimetry, Vol. 143, pp. 1713–1726, 2021.

[49] R. Ekiciler, K. Arslan, O. Turgut and B. Kursun, "Effect of hybrid nanofluid on heat transfer performance of parabolic trough solar collector receiver," Journal of Thermal Analysis and Calorimetry, Vol. 143, pp. 1637–1654, 2021.

[50] L. S. Sundar, A. H. Misganaw, M. K. Singh and A. Sousa, "Efficiency analysis of thermosyphon solar flat plate collector with low mass concentrations of ND–CO_3O_4 hybrid nanofluids: an experimental study," Journal of Thermal Analysis and Calorimetry, Vol. 143, pp.959-972, 2020.

[51] Z. Said, M. Ghodbane, L. S. Sundar, A. K. Tiwari, M. Sheikholeslami and B. Boumeddane, "Heat transfer, entropy generation, economic and environmental analyses of linear Fresnel reflector using novel rGO-Co_3O_4 hybrid nanofluids," Renewable Energy, Vol. 165, pp. 420-437, 2021.

[52] Y. Erdal and Y. Ali, "Assessments of thermal performance of hybrid and mono nanofluid U-tube solar collector system," Renewable Energy, Vol. 171, pp. 1079-1096, 2021.

[53] Z. Said, S. Arora and E. Bellos, "A review on performance and environmental effects of conventional and nanofluid-based thermal photovoltaics," Renewable and Sustainable Energy Reviews, Vol. 96, pp. 302–316, 2018.

[54] S. Saroha, T. Mittal, P. J. Modi, V. Bhalla, V. Khullar, H. Tyagi, R. A. Taylor and T. P. Otanicar, "Theoretical analysis and testing of nanofluids-based solar photovoltaic/thermal hybrid collector," Journal of Heat Transfer, Vol. 137, pp. 1-8, 2015.

[55] A. Ahmed, H. Baig, S. Sundaram and T. K. Mallick, "Use of Nanofluids in Solar PV/Thermal Systems," International Journal of Photo energy, Vol. 2019, pp. 1-17, 2019.

[56] M. Chen, Y. He, J. Huang and J. Zhu, "Synthesis and solar photo-thermal conversion of Au, Ag, and Au-Ag blended plasmonic nanoparticles," Energy Conversion and Management, Vol. 127, pp. 293-300, 2016.

[57] M. Chen, Y. He, J. Zhu and D. R. Kim, "Enhancement of photo-thermal conversion," Energy Conversion and Management, Vol. 112, pp. 21-30, 2016.

[58] Y. Khanjari, F. Pourfayaz and A. B. Kasaeian, "Numerical investigation on using of nanofluid in a water-cooled photovoltaic thermal system," Energy Conversion and Management, Vol. 122, pp. 263-278, 2016.

[59] Ali H. A. Al-Waeli, M. T. Chaichan, H. A. Kazem and K. Sopian, "Comparative study to use nano-(Al_2O_3, CuO, and SiC) with water to enhance Photovoltaic thermal PV/T collectors," Energy Conversion and Management, Vol. 148, pp. 963-973, 2017.

[60] J. Fang and Y. Xuan, "Investigation of optical absorption and photo thermal conversion characteristics of binary CuO/ZnO nanofluids," Royal Society of Chemistry, Vol. 7, pp. 56023-56033, 2017.

[61] S. Khashan, S. Dagher, S. Al Omari, N. Tit, E. Elnajjar, B. Mathew and A. H.-Alnaqbi, "Photo thermal characteristics of water-based Fe_3O_4@SiO_2 nanofluid for solar-thermal applications," Material Research Express, Vol. 4, pp. 1-10, 2017.

[62] M. S. Y. Ebaid, A. Ghrair, and M. Abd A. Al-Busoul, "Experimental investigation of cooling photovoltaic (PV) panels using (TiO_2) nanofluid in water-polyethylene glycol mixture and (Al_2O_3) nanofluid in water-cetyltrimethylammonium bromide mixture," Energy Conversion and Management, Vol. 155, pp. 324-343, 2018.

[63] A. Younis, M. Onsa, M., Y. Alhorr and E. Elsarrag, "The Influence of Al_2O_3- ZnO-H_2O Nanofluid on the Thermodynamic Performance of Photovoltaic- Thermal Hybrid Solar Collector System," Innovative Energy and Research, Vol. 7, pp. 1-8, 2018.

[64] N. S. Binti Rukman, A. Fudholi, N. F. Mohd Razali, M. H. Ruslan and K. Sopian, "Investigation of TiO_2 and MWCNT Nanofluids-based Photovoltaic-Thermal (PV/T) System," IOP Conference Series: Earth and Environmental Science, Vol. 268, pp.1-7, 2019.

[65] R. Sathyamurthy, A. E. Kabeel, A. Chamkha, A. Karthick, A. M. Manokar and M. G. Sumithra, "Experimental investigation on cooling the photovoltaic panel using hybrid nanofluids," Applied Nano Science, Vol. 11, pp. 363-374, 2021.

[66] I. W. Osho, H. Adun, M. Adedeji, E. C. Okonkwo, D. Kavaz and M. Dagbasi, "Effect of hybrid nanofluids mixture ratio on the performance of a photovoltaic thermal collector," International Journal of Energy Research, Vol. 44, pp. 1-18, 2020.

[67] H. Adun, M. Adedeji, M. Dagbasi, O. Bamisile, M. Senol and R. Kumar, "A numerical and exergy analysis of the effect of ternary nanofluid on performance of Photovoltaic thermal collector," Journal of Thermal Analysis and Calorimetry, Vol. 135, pp. 1-18, 2021.

[68] L. Sahota and G. N. Tiwari, "Effect of Al_2O_3 nanoparticles on the performance of passive double slope solar still," Solar Energy, Vol. 130, pp. 260-272, 2016.

[69] S. Elshamy and E. M. S. EI- Said, "Comparative Study based on Thermal, Exergetic and Economic Analyses of a Tubular Solar Still with Semi-circular Corrugated Absorber," Vol. 195, pp. 328-339, 2018.

[70] T.-X. He and L.-J. Yan, "Application of alternative energy integration technology in seawater desalination," Desalination, Vol. 249, pp. 104-108, 2009.

[71] B. Gupta, A. Kumar and P. V. Baredar, "Experimental Investigation on Modified Solar Still Using Nanoparticles and Water Sprinkler Attachment," Frontiers in Materials, Vol. 4, pp. 1-7, 2017.

[72] T. Elango, A. Kannan and K. Kalidasa Murugavel, "Performance study on single basin single slope solar still with different water nanofluids," Vol. 360, pp. 45-51, 2015.

[73] L. Sahota and G. N. Tiwari, "Effect of nanofluids on the performance of passive double slope solar still: A comparative study using characteristic curve," Desalination, Vol. 388, pp. 9-21, 2016.

[74] O. Mahian, A. Kianifar, S. Z. Heris, D. Wen, A. Z. Sahin and S. Wongwises, "Nanofluids effects on the evaporation rate in a solar still equipped with a heat exchanger," Nano Energy, Vol. 36, pp. 134-155, 2017.

[75] S. W. Sharshir, G. Peng, L. Wu, N. Yang, F. A. Essa, A. H. Elsheikh, S. I. T. Mohamed and A. E. Kabeel, "Enhancing the solar still performance using nanofluids and glass cover cooling: experimental study," Applied Thermal Engineering, Vol. 113, pp. 684-693, 2017.

[76] D. D. W. Rufuss, L. Suganthi, S. Iniyan and P. A. Davies, "Effects of nanoparticle-enhanced phase change material (NPCM) on solar still productivity," Journal of Cleaner Production, Vol. 192, pp. 9-29, 2018.

[77] W. Chen, C. Zou, X. Li and H. Liang, "Application of recoverable carbon nanotube nanofluids in solar desalination system: An experimental investigation," Desalination, Vol. 451, pp. 92-101, 2019.

[78] H. Md. F. Rabbi and A. Z. Sahin, "Performance improvement of solar still by using hybrid nanofluids," Journal of Thermal Analysis and Calorimetry, Vol. 143, pp. 1345-1360, 2021.

[79] S. M. S. Murshed, C. A. N. De Castro, M. J. V. Lourenco, M. L. M. Lopes and F. J. V. Santos, "A review of boiling and convective heat transfer with nanofluids," Renewable and Sustainable Energy Reviews, Vol. 15, pp. 2342-2354, 2011.

7

Modification and Application of Vegetable Oils for Biofuels

Madhu Agrawal and Neha Pal

Department of Chemical Engineering, Malaviya National Institute of Technology Jaipur 302017, India
E-mail: madhunaresh@gmail.com

Abstract

Vegetable oils are an environmentally friendly renewable resource that contains a wide range of fatty acid compositions because this is suitable for various industrial purposes. The products derived from vegetable oil are non-toxic, environmentally friendly, and biodegradable. So, this is an attractive natural source for developing new renewable energy. The depletion of fossil fuels attract researcher to innovate new renewable and environmentally friendly energy source. The biodiesel derived from vegetable oil is a promising energy source that can replace fossil fuels. This chapter comprises an overview of vegetable oil feedstock, the effect of direct use of vegetable oil in diesel engines, biodiesel derived from vegetable oil, and the current status of biodiesel in India.

Keywords: Vegetable oil, Biodiesel, Fatty acid, Transesterification.

7.1 Introduction

The vegetable oils are obtained from various types of oilseed crops which are commonly used for food purposes. Mainly vegetable oil is used for edible purposes, but its characteristics make it a valuable material for industrial applications. Vegetable oil contains a wide variety of fatty acids with various configurations which give them specific physio-chemical characteristics [1]. A fatty acid is essential for various industrial purposes such as biodiesel production, soaps and detergents, varnishes, paints, cosmetics, lubricants, inks, and so on.

Therefore, the demand for oilseed crops increases day by day. Typically, vegetable oil contains Fatty acids that can carry zero to three double bonds. The fatty acid mainly found in vegetable oil are saturated fatty acid (arachidic, palmitic, stearic, lignoceric, myristic), unsaturated fatty acid (petroselinic, oleic, palmitoleic, gadoleic) and polyunsaturated fatty acid (eleostearic, linoleic, linolenic, calendic) [2]. In saturated fatty acids since all carbon atoms are bonded to hydrogen (H2), oxygen (O2), and carbon (C) molecules, therefore no carbon doubles (C-C) bound. In unsaturated fatty acid, when a couple of hydrogen atoms are removed from a fatty acid, one double bond is present in it and the remaining carbon atom is in a single bound. In polyunsaturated fatty acids, more than one unsaturated fatty acid is present. These types of fatty acids are primarily stored in the form of triacylglycerol (TAG) which is the major storage medium seed. The vegetable oil used as a feedstock has several advantages: 1) renewable source, 2) substrate generated by the industry are toxic-free or less toxic, 3) products developed are biodegradable, and 4) environmentally friendly products generated [3, 4].

Among the various industrial application of vegetable oil, biodiesel production is very attractive nowadays. For biodiesel, lipids-based feedstocks are required, and vegetable oils are a rich and economical source of lipids. Edible and non-edible are both types of vegetable oil suitable for the production of biodiesel. The major variation between the different kinds of vegetable oils is the variety of fatty acids bound with triglycerides molecules. Table 7.1 shows the composition of fatty acids present in different vegetable oils. The composition of fatty acids present in vegetable oil is an important property as it determines the biodiesel property [2]. Table 7.3 shows a different kind of fatty acid present in vegetable oil. The presence of fatty acids influences the molecular weight and degree of unsaturated/saturated vegetable oil. The degree of unsaturated/saturated and molecular weight can be determined by saponification value and iodine value.

7.2 History of Vegetable Oil as a Fuel

The direct use of vegetable oil in a diesel engine was investigated various years ago by Rudolf Diesel. Rudolf Diesel, inventor of the diesel engine first time tested peanut oil as a fuel directly in the diesel engine. This engine was

Table 7.1 Fatty acid composition of various vegetable oil

Seed plants	Oilseed price (USD/ton)	Oil content (%)	Fatty acid				Reference
			Oleic (%)	Linoleic (%)	Stearic (%)	Palmitic (%)	
Soybean	684	21	25	52	4	11	[5]
Sunflower	46	46	46.02 ± 16.75	0.12 ± 0.09	45.39 ± 18.77	1.98 ± 1.44	[2]
Palm	478	40	41.90 ± 1.20	11.03 ± .02	3.39± 0.65	41.78 ± 1.27	[2]
Coconut	812	60-63	5.84 ± 0.50	1.28 ± 0.18	1.94 ± 0.17	8.62 ± 0.50	[2]
Jatropha	–	30-40	42.4±6.2	28.8±7.6	13.25±1.25	7±0.7	[6]
Peanut	1253	51.9	43.2	27±5	–	11±3	[2]
Canola	683	48.3	56.80–64.92	17.11–20.92	–	4.18–5.01	[7]
Cottonseed	–	15–20	19.2	55.2	–	20.1	[8]
Rice bran oil	–	16–32	42	32	–	20	[9, 10]

initially designed for diesel fuel, without any modification of vegetable oil used in the engine, it gives satisfactory performance. In ancient times, various types of vegetable oil used by the researcher directly in a diesel engine are palm oil, castor oil, soybean oil, cottonseed oil, etc. [11].

During World War II, Vegetable oils were used for driving diesel engines in remote locations such as Abidjan Port (Ivory Coast) [12, 13]. In the Ivory Coast supply of fuel is very difficult. So port Construction Company used filter palm oil for operating 50–800 hp engines. Almeida et al. tested that the actual fuel usage and temperature of the exhaust gas increased as higher loads enhanced the CO emissions and the percentage of charge increased. It was because there was a lack of oxygen at low equivalence ratios. NOx Emissions of palm oil were lower than diesel [14]. Masjuki et al. (2000), worked on a similar study preheated palm oil was directly injected into a diesel engine. The fuel viscosity decreased during preheating, and the effect of spraying and atomization was also observed. Torque, fuel consumption, exhausts, thermal brake, and brake strength performance were found to be similar to diesel fuel [15]. Abbass et al. (1990), tested pure sunflower oil in diesel engines and observed high emissions of hydrocarbon, NOx, and CO compared with diesel fuel due to shorter ignition delays and higher diffusive burns [16]. Dhingra et al. (1993), Karanja oil, rice bran oil, and neem oil were tested on a low heat-rejection engine. They used an electric heater to preheat the vegetable oil. Without heating, the performance was stated to be 1–4% lower than that of mineral diesel. It however improved with vegetable oils being preheated [17].

Many researchers work on blended vegetable oil with diesel fuel to reduce the cost of conventional fuels. Wang et. al. (2006), Vegetable oil blended with diesel. They observed higher exhaust gas temperatures as compared to diesel fuel, with very limited CO variations and low NOx [18].

So many works conducted on different vegetable oil directly used in a diesel engine but the big downside of vegetable oils used in a diesel engine is their high viscosity, caused by a polymerization reaction that produces large molecular weight components which creates trumpet and coking on the injectors resulting in low atomization and eventually leads to technical challenges such as carbon deposition in an engine [19].

7.3 Transesterification of Vegetable Oil

Transesterification is the widely used and well-accepted chemical reaction where fatty acids (vegetable oil) and alcohol react in the presence of catalysts to produce biodiesel and glycerol. The key element of vegetable oils is triacyl-glycerol, a molecule comprised of three esters of the fatty acid chain bound to a backbone of glycerol (Figure 7.1) (Issariyakul and Dalai, 2014). Replacement of one or two acyl groups with the hydroxyl groups (–OH) is called monoacylglycerol (MAG) and diacylglycerol (DAG), respectively The reaction is effect by various factors 1) oil to alcohol ratio, 2) reaction time, 3) nature of feedstock composition (vegetable oil), and type and amount of catalyst [20].

Alcohol is a very important parameter in the transesterification reaction. Various type of alcohol has been investigated for biodiesel processing. The most commonly used acyl accepter alcohol is methanol and to a slight extent, ethanol due to their physical and chemical property (short-chain alcohol and polar) and low price. Isopropanol, octanol, propanol, and branched alcohol are also used but these alcohols are costly (Musa, 2016). The presence of catalyst significantly accelerated the reaction. Homogenous and heterogeneous

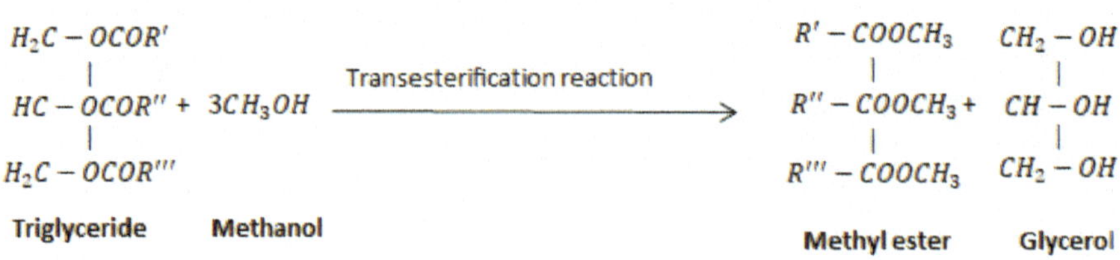

Figure 7.1 Transesterification reaction

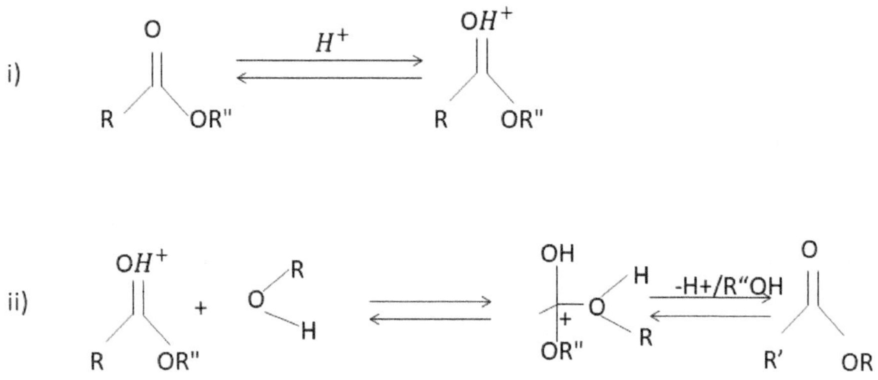

R"= Hydrogen for transesterification
R'= Carbon chain of fatty acid
R= Alkyl group of alcohol

Figure 7.2 Mechanism of homogeneous catalyst in transesterification reaction [2].

are both types of catalysts used for biodiesel production. Homogeneous catalysts most popular include alkalis and acids. NaOH, KOH, and related sodium and potassium alkoxides are the most widely used alkali catalysts. In the acid-catalyzed reaction sulphuric acid (H2SO4), hydrochloric acid (HCl), and sulfonic acid (H2O3S) are typically used as catalysts. Mechanism of homogenous catalyst in transesterification reaction shown in Figure 7.2. The use of a homogeneous catalyst for biodiesel production has some disadvantages such as water formation during acid esterification and its corrosive nature Heterogeneous acid catalyst are highly recommended for transesterification reactions because they are easily separated and have less waste generation. The common heterogeneous catalysts are zeolites, sponsored alkali metals, Alkaline earth metals, and hydrotalcite [21].

7.4 Biodiesel Feedstock

Various types of vegetable oil such as edible oil (sunflower, Soybean, Palm, and Rapeseed) and non-edible oil (Jatropha, Rice bran oil,) are used for biodiesel production using the transesterification process. Table 7.2 describes the physical property of the vegetable used.

Table 7.2 Physical or thermal property of vegetable oil [19]

Vegetable oil	Kinematic Viscosity (40 °C)	Cetane no	Density (Kg/l)	Carbon residue (wt. %)	Heating value (MJ/kg)	Pour point (°C)	Cloud point (°C)	Flash point (°C)
Soybean	32.6	37.9	0.9138	0.27	39.6	-12.2	-3.9	254
Sunflower	33.9	37.1	0.9161	0.23	39.6	-15	7.2	274
Palm	39.6	42	0.9180	–	–	–	31	267
Jatropha	35.98	45	0.9186	0.44-0.64	39	3-5	8-10	225-233
Peanut	39.6	41.8	0.9026	0.24	49.8	-6.7	12.8	271
Rapeseed	37	37.6	0.9115	0.30	39.7	-31.7	-3.9	246
Cottonseed	33.5	41.8	0.9148	0.24	39.5	-15	1.7	234

Table 7.3 Common fatty acid found in vegetable oil [2]

Fatty acid	Formula	Symbol
Saturated fatty acid		
Palmitic	$C_{16}H_{32}O_2$	C16:0
Stearic	$C_{18}H_{36}O_2$	C18:0
Lignoceric	$C_{24}H_{48}O_2$	C24:0
Myristic	$C_{14}H_{28}O_2$	C14:0
Arachidic	$C_{20}H_{40}O_2$	C20:0
Capric	$C_{10}H_{20}O_2$	C10:0
Unsaturated fatty acid		
Oleic	$C_{18}H_{34}O_2$	C18:1
Palmitoleic	$C_{16}H_{30}O_2$	C16:1
Gadoleic	$C_{20}H_{38}O_2$	C20:1
Petroselinic	$C_{18}H_{34}O_2$	C18:1
Polyunsaturated		
Linoleic	$C_{18}H_{32}O_2$	C18:2
Linolenic	$C_{18}H_{30}O_2$	C18:3
Calendic	$C_{18}H_{30}O_2$	C18:3
Eleostearic	$C_{18}H_{30}O_2$	C18:3

7.4.1 Palm Oil

Africa is the world largest producer of palm oil. Southeast Asia's producing areas are, particularly Indonesia and Malaysia, which together contribute to developed palm oil around 80% of the overall world production. India is the world's 17th-largest producer of palm oil. There have been generally two types of palm oil namely mesocarp palm oil and kernel palm oil inside the crop. Palm oil contains more saturated fatty acid than soybean oil and rapeseed oil. The main fatty acid present in palm oil is linoleic, oleic, palmitic, and stearic are the main fatty acids.

The biodiesel production using palm oil is a promising feedstock among the others due to its physical and chemical characteristics. Palm oil derives biodiesel has various advantages such as low sulphur content, lower pour point due to this easily used in a cold climate, fewer carbon residues, and high oxidation value [22]. Due to these advantages, many works are done on it Soetaredjo et al. (2011), prepared biodiesel from palm oil using serious KOH/bentonite catalyst via transesterification process [23]. Madhuvilakku and Piraman et al., also developed biodiesel by transesterification using mixed oxide nanocatalyst (Zno-TiO2) [24]. Many other researchers work on palm oil using different types of a catalyst such as KOH/Al2O3 and KOH/NaY [25], Rice husk based catalyst [26], and KF/Hydrotalcity [27], Cao-CeO2 mixed oxide catalyst [28]. Palm oil-based biofuel is a potential fuel for combustion engines but it suffers some problems such as high viscosity, a high flashpoint, and low gross heat of combustion [22].

7.4.2 Sunflower Oil

Sunflower botanical name is genus Helianthus annuus that belongs to the Compositae family of flowering plants that grows all over the world. The name is derived from the Greek words helios (sun) and anthos (herb). The highest production of sunflower is found in Mexico and the Southwest United States. India is in 18th position according to the worldwide sunflower production. The major states of India for sunflower production are Tamil Nadu, Orissa, Maharashtra, Bihar, and Andhra Pradesh is major sunflowers producing states of India. Sunflower seeds are nutritious and are often broken to extract oil. Oleic acid, Stearic acid, Palmitic acid, and linoleic acid are the main fatty acids present in sunflower oil. The sunflower is one of the most ancient oleaginous plants, as it can be cultivated Built from three thousand B.C. Before the emergence of the soybean booming after World War II, the Sunflower was once the world's top oil-producing plant. It contained useful fatty acid which is responsible for biodiesel production and high oil contains motivated researchers used as a feedstock for biodiesel synthesis. Agarwal et al. (2012), used the sunflower as a feedstock and converted it into biodiesel via transesterification reaction using KOH catalyst and methanol as alcohol [29]. A similar study by Requena et al. (2011), also prepared biodiesel using methanol as alcohol and NaOH as a catalyst. Many works are done by the researcher on biodiesel production from edible oil but the main drawback of edible oil is affecting the food chain [30].

7.4.3 Soybean Oil

"Soybean" related to the Glycine max is as that is found only under production and is a part of the Papilionaceae. The source of soybean is not clear, because there is two significant gene center for the genus Glycine; East Africa and Australia. The genus Glycine is thought to have been spread by migratory birds as carriers of seeds from Australia to the entire Pacific region including China. Soybean first was discussed in the literature of the United States in 1804, used primarily as a forage crop, which was not a significant crop until around the Second World War [31]. Its production and economy have developed exponentially in the USA. Nowadays, India is the 5th largest producer of soybean seeds, and Rajasthan, Maharashtra, and Madhya Pradesh together contribute to around 92–93% of the production of soybean in India.

The several advantages that using soybean oil in biodiesel production include soybeans being widely grown, the infrastructure and equipment already existing for growing, transporting, and processing and the leftover soybean meal being useful for animal feed. Due to these advantages, Gupta et al. (2018), produce biodiesel via transesterification using marble slurry derived calcined marble slurry (CMS), and hydroxyapatite (HAP) as a heterogeneous catalyst. Results show that HAP provides better yield as compared to CMS catalyst. The highest yield of biodiesels is 94% under optimum conditions (65 °C, reaction time 3h, methanol to oil ratiot9:1) [32]. The challenge of soybean oil and other edible for the production of biodiesel is an effect on food vs. fuel. Soya oil is often used in nutrition products for human use, as cooking oil, and for many industrial applications. Soybeans account for 80% or more of India's edible fats and oils. Competition with other uses has triggered soybean oil market price increases, threatening soybean biodiesel profitability.

7.4.4 Rapeseed Oil/Canola Oil

The word "rape" has its roots in the Latin word "rapum," meaning turnip. Rapeseed oil/canola oil is a member of the Brassica family comprising mustard, broccoli, cod, and rutabagas. Brassica crops can survive in cold weather conditions. Sinapis include the strategically valuable crops of the species. Brassica is of many types such as Sinapis Arvensis (wild mustard), Brassica nigra (black mustard), Brassica juncea (brown, oriental), Sinapis alba (white mustard), Brassica oleracea (broccoli, kale, cauliflower), Brassica campestris (turnip rape), Brassica carinata (Abyssinian mustard), and Brassica juncea (leaf mustard, brown and oriental). These seeds contain oil of around 40% of the major fatty acids including erucic acid (C22:1) oleic acid (C18:1), and linoleic acid (C18:2) [33].

Rapeseed/canola oil contained a high oil percentage as compared to other vegetable oil. This oil has very little saturated fat (7%) this biodiesel produces from rapeseed/canola oil easily performs in cold weather. Due to its benefit for biodiesel production Wang et al. (2013), worked on the trans-esterification of rapeseed using methanol as alcohol. For enhanced reaction, a solid base catalyst (Ca12Al14O33 and CaO) was utilized. The outcome of the reaction provides a 90% conversion of oil to biofuel [34]. In a similar study by Georgogianni et al. (2009) using homogeneous (NaOH) or heterogeneous (Mg-MCM-41, Mg-Al Hydrotalcite, and K+ impregnated zirconia) the catalyst in transesterification reaction for the production of biodiesel. It concludes that heterogeneous catalysts perform better than homogeneous catalyst. Results show that Mg-Al Hydrotalcite catalysts provide the highest yield as a camper to another heterogeneous catalyst. The biodiesel yield using Mg-Al Hydrotalcite catalyst in transesterification reaction is 96% [35]. The Heterogeneous solid base catalysts can be easily separated from the reaction mixture by simple filtration; they are easily recovered and carry a less corrosive quality, resulting in safer, simpler, and more environmentally friendly processes (Georgogianni et al., 2009).

7.4.5 Rice Bran Oil

Rice bran oil is obtained from the outer surface of rice grain and it is non-edible oil. India is the first largest producer of rice bran oils its production around tonnes/annum. Rice bran contains around 16–32% oil content which depends upon rice verity and degree of milling (Mazaheri et al., 2017). Major fatty acid composition in rice bran oil is oleic acid (42%), linoleic acid (32%), and palmitic acids (20%) [36]. Free fatty acid present in rice bran oil is similar to edible oil used for the production of biodiesel. Thus, it makes a suitable feedstock for biodiesel production.

Rice bran is an agricultural waste that is produced a huge amount in India as well as it contains various types of useful fatty acids which responsible for the production of biodiesel. Due to these advantages, various researchers work on rice bran oil using different catalysts. Mazaheri et al., (2017) used a CaO catalyst for the conversion of rice bran oil into biodiesel using a transesterification reaction. The results show that the highest yield comes at optimum parameters (methanol to oil ratio: (35:1); reaction time: 95 min;

reaction temperature: 1100 °C: catalyst weight: 0.5g). Rice bran oil provides high-quality biofuel even then it is not commercialized due to its drawbacks such as Pre-treatment of oil is needed, high conversion not archived at a short interval of time, various by-products (defatted rice bran) generated during the production of biodiesel [37].

7.4.6 Jatropha

The Euphorbiaceae family includes the Jatropha circus. It originated in America, but it is now mostly grown in Asian countries, especially India. Jatropha is adapted to dry and semi-dry climates and it scrapes the leaves to survive during droughts. This is one of the most essential feedstock for the production of biofuels since it may be produced on uncultivated and damaged wasteland. While the Jatropha plant has lower nutrient requirements, the production of Jatropha in acid soils requires extra minerals such as vitamins and magnesium, because of its alkaline soil preference [38]. Jatropha-derived oil is nonedible because of the cursing, a poisonous compound found in plants. The major fatty acid present in jatropha is oleic acid (43–49%), linoleic acid (29–35%), palmitic (13–15), and stearic acid (7–8%) [6]. Low alkalinity, superior corrosion resistance compared to soya oil, low viscosity related to castor oil, and higher cooling qualities related to palm oil make this the most impressive result. Even during the storage cycle, the viscosity, free fatty acids, and oil/biodiesel density are also steady. These properties make the jatropha plant a promising candidate for biodiesel production [39]. Various types of catalysts are used for the production of biodiesel such as KOH [29], solid catalyst CaO [40], metal mixed oxide catalyst (TiO2-ZnO) [24], homogeneous (NaOH), and heterogeneous (Na2ZrO3) catalyst [41], and mixed oxide catalyst (CaO–La2O3) [42]. Taufiq-Yap et al. (2014) stated that the co-existence of two separate basic oxides (CaO–La2O3) in the catalyst could boost the catalyst efficiency and be appropriate for transesterification reactions. It provide 86.1% biodiesel conversion [42]. Nevertheless, limited research to generate biodiesel using bimetallic oxides has been published in the literature. In similar work of Teo et al. (2015), used mixed metal oxide (CaLa2O3) catalyst for biodiesel production from jatropha oil. The result shows more than 80% biodiesel yield [43]. Many researchers developed biodiesel from jatropha oil using different kinds of catalysts for enhanced the yield of biodiesel. But nowadays jatropha oil needed more technical enhancement for increasing the biodiesel production yield.

7.4.7 Used Cooking Oil

The characteristic of used-cooking oil is strongly dependent upon the oil's origin and history. The source of used cooking oil defines the composition of fatty acids, the physicochemical characteristics of the oil i.e. free fatty acid content, viscosity, consists of polymerized oil, water content, and consists of oxidised oil compounds. During cooking, vegetable oil is degraded by three main reactions: 1) oxidation, 2) thermolytic, and 3) hydrolytic. Saturated fatty acids must be a breakdown at a high temperature to generate ketones, alkanes, diacylglycerides, fatty acids, esters, and other compounds [2].

The cost of vegetable oil key factor for the production of biodiesel and edible oils directly affected for food chain due to this cost of vegetable oil may be increased. Waste cooking oil is a promising alternative to low-cost oil for biofuel production. Due to its low-cost Agarwal et al. (2012) synthesis biodiesel from waste cooking oil via transesterification reaction using KOH loaded with alumina catalyst [29]. Many other researchers work on using cooking oil at different catalysts using the transesterification process such as Patil et al., (2010) used ferric sulphate [44], Tan et al. (2019), used chicken and fish bones catalyst. Chicken and fish bones catalyst very efficiently converted waste cooking oil to biodiesel. The conversion efficiency is 89.5% at optimum parameter (temperature 65 °C; catalyst weight 1.65w/v%; reaction time1.54 h) [45]. Farooq and Ramli, (2015) used heterogeneous catalysts derived from waste chicken bones and it provide an 89.33% biodiesel conversion rate [46]. Heterogeneous catalysts are very effective for using cooking oil conversation into biodiesel because these types of catalysts facilitate fast recovery, recyclability, and a green process that is cost-efficient. Such catalysts withstand high free fatty acid and a large amount of moisture. Effective and cheap heterogeneous catalysts help reduce the overall cost of processing biodiesel. Using waste cooking oil is an economically feasible material for biodiesel production. The main problem that arises with waste cooking oil is the high percentage of fatty acid due to this alkaline catalyst reacting with fatty acid and forming soap and resultant biodiesel production effected [5]. To reduce the fatty acid contents pre-treatment is needed. Gnanaprakasam et al. (2013), successfully reduce the fatty retreatment with an acid catalyst to undergo an esterification reaction, which needs high operating conditions [47]. Ho et al. (2014) also work on the reduction of fatty acids by catalyzed waste cooking oil using a homogeneous acid catalyst (SrFe2O4/SiO2-SO3H). This catalyst almost 100% converted waste cooking oil into biodiesel [48].

Various researchers used different types of catalysts for biodiesel production, whenever biodiesel yield and technology used for conversion are still challenging.

7.5 Biodiesel

The demand for energy increases due to a reaped increase in population and urbanization [49]. According to IEA, The oil demand used in passenger vehicles peaked in the late 2020s, and during the 2030s this demand increased by 0.1 Mb/d on average per year. Nevertheless, there is no definite peak in oil use, because automobile companies, aviation sectors, and the shipping sector increase continuously. Based on the current energy consumption we need an alternative source to fulfill the energy requirements. Fossil fuel-based energy sources create global warming and climate change due to greenhouse gas emissions [50]. Among all alternative renewable energy sources (nuclear energy, solar, water, and wind), green energy source (biofuel) are more attractive because these sources can be used as fuel and in other value-added chemicals [5], it is easily biodegradable and less toxicity, this seems to have greatly increased the economic prospects of rural areas, and due to the closed carbon cycle does not lead to global warming [51]. Additionally, Biofuels are environmentally friendly, resulting in a very small release of sculptures and no net accelerated release of, aromatic compounds carbon dioxide, or other environmentally damaging chemical substances [51, 52]. The vegetable oil is divided into two parts: 1) edible oil, and 2) non-edible oil.

Biodiesel production from edible oils such as rapeseed, soybeans, palm oil, sunflower oil, etc. It uses as a blend with traditional fuels (10% ethanol) and it can be used in various vehicle technology such as FFVs (Flexible Fuel Vehicle) or natural gas vehicles [52]. The main disadvantage of this generation are 1) imbalance in the global food market due to this world can face a "food vs. fuel" crisis [50, 52] a large portion of land required for cultivation to fulfill the demand for biofuel due to this serious ecological imbalance creates as began cutting down the forest for farming purpose [50]. Hence, edible feedstock, is not cost-efficient, these feedstocks could cause deforestation. Additionally concerned about the availability of land, water for cultivation, and fertilizer use in the cultivation of lignicolous based plants [53, 54]. To reduce the dependency on edible oil need an alternative feedstock that reduces these drawbacks .therefore, move to non-edible vegetable oil feedstock.

Biodiesel is synthesized from a non-edible vegetable oil source. The energy crop is divided into two categories 1) the Grassy (herbaceous or forage), and 2) woody (tree) energy crops. The most common energy vegetable oil for this purpose is jatropha, rice bran oil, salmon oil, Rubber tree, and jojoba oil. Besides, used cooking oils also are utilized as biodiesel feedstock. The biofuels derived from non-edible vegetable oil avoid the competition of food vs. fuel which directly impacts food prices. Although these biofuels are currently being researched in terms of conversion technology and process operation, once commercialized, they should meet needs for minimal land usage and substantially improved CO2 emission reduction potential. According to the published literature, the feedstock costs contribute to 46–59% of overall expenses for biofuels obtained from non-edible oil, depending on conversion efficiency and used technology.

7.6 The Current Senior of Biodiesel Derive from Vegetable Oil in India

India has long been using bio-energy as a source of energy directly for household cooking through the use of wood and other bio-waste plants, as biomass combustion now accounts for 24% of India's energy demand [8]. However, there has been no substantial improvement in the use of biodiesel such as vegetable oils, ethanol, butanol, etc. Such fuels were used extensively for current research but have yet to take off their industrial production to meet the oil demand. As a consequence, the present biofuels rates are significantly higher. The Indian government has been affording two initiatives to increase the productivity of these biofuels, one being ethanol-blended petrol and the other being biodiesel production. In India, the primary sources of biodiesel production are used cooking oils, vegetable oil (edible and non-edible oil), and animal fats. The development of biodiesel by seed oil such as (oil obtained from non-edible). Currently, oil produced from jatropha seeds is a major research field in India.

Under the ethanol-blended petrol scheme, the government has ordered oil companies to provide ethanol-blend gasoline with 10% ethanol content. The use of ethanol-blended petrol was encouraged in 2014. The cost of ethanol has been established by the government at Rs. 48.50–49.50 per litre [55]. India's total production capacity for ethanol stands at 4.5 billion liters. The estimated mixing rate for unit corporations in the public sector is 3.56%. The Indian government introduced the purchase policy for biodiesel in 2005 and, on 16.01.2015, the government allowed the producer to directly sell biodiesel B100 to oil manufacturing companies. Biodiesel (B100) was produced in the second quarter of 2016 at 33.7 million liters. Nowadays biodiesel blended diesel sell in India [56, 57]. Jatropha is non-edible oil that can develop in adverse weather conditions may be advantageous in India a country where environmental conditions vary across the region. India's vast agricultural acreage might be used to generate biodiesel feedstock like jatropha, and its growth could be industrialized in the future.

7.7 Conclusion

The green and environmentally friendly energy sources are expected to be in high demand as traditional fuels decline very drastically. Vegetable oils are the source of a variety of fatty acids, therefore, they can be used in a diesel engine as a fuel but they created large carbon residue due to this engine no longer being working. Therefore, the use of biodiesel derived from vegetable oils would play an important role, especially in the transport sector. Various factors affected biodiesel production such as the type of vegetable oil used as feedstock, the manufacturing process, and the cost of oil. Vegetable oil is different from each other due to the presence of a variety of fatty acid in different compositions this biodiesel derived from these vegetable oils have different characteristics. The edible oils such as palm oil having high oxidation stability, and rapeseed /canola oil are promising oil for cold climates due to its property making promising components but edible oil directly impacts conflicts with food supply. Therefore non-edible (jatropha) sources are more in demand because this type of vegetable oil avoids the competition of fuel vs. fuel. The non-edible vegetable oil contained a high amount of fatty acid which increases the biodiesel production cost. So, these fuels are also under technology investigation. Various types of catalysts are used for the production of biodiesel for enhancing the yield of biodiesel production and to improve in technology. Among the different catalysts heterogeneous solid base catalysts are highly effective due to their advantages such as 1) being easily separated from the reaction mixture, 2) very less corrosive nature, and 3) environmentally friendly processes.

References

[1] Tabatabaei M, Aghbashlo M, Dehhaghi M, Panahi HKS, Mollahosseini A, Hosseini M, et al. Reactor technologies for biodiesel production

and processing: A review. Prog Energy Combust Sci 2019;74:239–303. https://doi.org/10.1016/j.pecs.2019.06.001.

[2] Issariyakul T, Dalai AK. Biodiesel from vegetable oils. Renew Sustain Energy Rev 2014;31:446–71. https://doi.org/10.1016/j.rser.2013.11.001.

[3] Viccaro M, Cozzi M, Rocchi B, Romano S. Conservation agriculture to promote inland biofuel production in Italy: An economic assessment of rapeseed straight vegetable oil as a self-supply agricultural biofuel. J Clean Prod 2019;217:153–61. https://doi.org/10.1016/j.jclepro.2019.01.251.

[4] Rathore V, Newalkar BL, Badoni RP. Processing of vegetable oil for biofuel production through conventional and non-conventional routes. Energy Sustain Dev 2016;31:24–49. https://doi.org/10.1016/j.esd.2015.11.003.

[5] Rodionova M V., Poudyal RS, Tiwari I, Voloshin RA, Zharmukhamedov SK, Nam HG, et al. Biofuel production: Challenges and opportunities. Int J Hydrogen Energy 2017;42:8450–61. https://doi.org/10.1016/j.ijhydene.2016.11.125.

[6] Adewuyi A, Adebowale KO. Addition of an azido group to epoxidised fatty acid methyl esters of Jatropha curcas by epoxy cleavage. Res Chem Intermed 2014;40:2015–23. https://doi.org/10.1007/s11164-013-1099-y.

[7] Matthaus B, Özcan MM, Juhaimi F Al. Some rape/canola seed oils: Fatty acid composition and tocopherols. Zeitschrift Fur Naturforsch - Sect C J Biosci 2016;71:73–7. https://doi.org/10.1515/znc-2016-0003.

[8] Griswold MG, Fullman N, Hawley C, Arian N, Zimsen SRM, Tymeson HD, et al. Alcohol use and burden for 195 countries and territories, 1990-2016: A systematic analysis for the Global Burden of Disease Study 2016. Lancet 2018;392:1015–35. https://doi.org/10.1016/S0140-6736(18)31310-2.

[9] Ju YH, Vali SR. Rice bran oil as a potential resource for biodiesel: A review. J Sci Ind Res (India) 2005;64:866–82.

[10] Mazaheri H, Ong HC, Masjuki HH, Amini Z, Harrison MD, Wang CT, et al. Rice bran oil based biodiesel production using calcium oxide catalyst derived from Chicoreus brunneus shell. Energy 2018;144:10–9. https://doi.org/10.1016/j.energy.2017.11.073.

[11] Knothe G. History of Vegetable Oil-Based Diesel Fuels. Second Edi. AOCS Press; 1913. https://doi.org/10.1016/B978-1-893997-62-2.50007-3.

[12] Demirbas A. Biodiesel from waste cooking oil via base-catalytic and supercritical methanol transesterification. Energy Convers Manag 2009;50:923–7. https://doi.org/10.1016/j.enconman.2008.12.023.

[13] Gnanasekaran D, Chavidi VP. Vegetable Oil: An Eco-friendly Liquid Insulator 2018:101–24. https://doi.org/10.1007/978-981-10-4870-8_6.

[14] Petrenko C, Paltseva J, Searle S. Ecological impacts of palm oil expansion in Indonesia |International Council on Clean Transportation. Washingt Int Counc Clean Transp 2016:1–21.

[15] Maleque MA, Masjuki HH, Haseeb ASMA. Effect of mechanical factors on tribological properties of palm oil methyl ester blended lubricant. Wear 2000;239:117–25. https://doi.org/10.1016/S0043-1648(00)00319-7.

[16] Abbass MK, Andrews GE, Williams PT, Bartle KD, Davies IL, Lalah JO. The composition of the organic fraction of particulate emissions of a diesel operated on vegetable oil. SAE Tech Pap 1990. https://doi.org/10.4271/901563.

[17] Dhinagar SJ, Nagalingam B, Gopalakrishnan K V. Experimental investigation of non-edible vegetable oil operation in a LHR diesel engine for improved performance. SAE Tech Pap 1993. https://doi.org/10.4271/93 2846.

[18] Wang YD, Al-Shemmeri T, Eames P, McMullan J, Hewitt N, Huang Y, et al. An experimental investigation of the performance and gaseous exhaust emissions of a diesel engine using blends of a vegetable oil. Appl Therm Eng 2006;26:1684–91. https://doi.org/10.1016/j.applther maleng.2005.11.013.

[19] Capuano D, Costa M, Di Fraia S, Massarotti N, Vanoli L. Direct use of waste vegetable oil in internal combustion engines. Renew Sustain Energy Rev 2017;69:759–70. https://doi.org/10.1016/j.rser.2016.11.0 16.

[20] Keera ST, El Sabagh SM, Taman AR. Transesterification of vegetable oil to biodiesel fuel using alkaline catalyst. Fuel 2011;90:42–7. https: //doi.org/10.1016/j.fuel.2010.07.046.

[21] Gupta D, Kumar R, Pant KK. Hydrotalcite supported bimetallic (Ni-Cu) catalyst: A smart choice for one-pot conversion of biomass-derived platform chemicals to hydrogenated biofuels. Fuel 2020;277:118111. ht tps://doi.org/10.1016/j.fuel.2020.118111.

[22] Ludin NA, Bakri MAM, Kamaruddin N, Sopian K, Deraman MS, Hamid NH, et al. Malaysian oil palm plantation sector:

Exploiting renewable energy toward sustainability production. J Clean Prod 2014;65:9–15. https://doi.org/10.1016/j.jclepro.2013.11.063.

[23] Soetaredjo FE, Ayucitra A, Ismadji S, Maukar AL. KOH/bentonite catalysts for transesterification of palm oil to biodiesel. Appl Clay Sci 2011;53:341–6. https://doi.org/10.1016/j.clay.2010.12.018.

[24] Madhuvilakku R, Piraman S. Biodiesel synthesis by TiO2-ZnO mixed oxide nanocatalyst catalyzed palm oil transesterification process. Bioresour Technol 2013;150:55–9. https://doi.org/10.1016/j.biortech.2013.09.087.

[25] Noiroj K, Intarapong P, Luengnaruemitchai A, Jai-In S. A comparative study of KOH/Al2O3 and KOH/NaY catalysts for biodiesel production via transesterification from palm oil. Renew Energy 2009;34:1145–50. https://doi.org/10.1016/j.renene.2008.06.015.

[26] Chen GY, Shan R, Shi JF, Yan BB. Transesterification of palm oil to biodiesel using rice husk ash-based catalysts. Fuel Process Technol 2015;133:8–13. https://doi.org/10.1016/j.fuproc.2015.01.005.

[27] Gao L, Xu B, Xiao G, Lv J. Transesterification of palm oil with methanol to biodiesel over a KF/hydrotalcite solid catalyst. Energy and Fuels 2008;22:3531–5. https://doi.org/10.1021/ef800340w.

[28] Lee SL, Wong YC, Tan YP, Yew SY. Transesterification of palm oil to biodiesel by using waste obtuse horn shell-derived CaO catalyst. Energy Convers Manag 2015;93:282–8. https://doi.org/10.1016/j.enconman.2014.12.067.

[29] Agarwal M, Chauhan G, Chaurasia SP, Singh K. Study of catalytic behavior of KOH as homogeneous and heterogeneous catalyst for biodiesel production. J Taiwan Inst Chem Eng 2012;43:89–94. https://doi.org/10.1016/j.jtice.2011.06.003.

[30] Iglesias L, Laca A, Herrero M, Díaz M. A life cycle assessment comparison between centralized and decentralized biodiesel production from raw sunflower oil and waste cooking oils. J Clean Prod 2012;37:162–71. https://doi.org/10.1016/j.jclepro.2012.07.002.

[31] Zheng Z, Wang J, Wei Y, Liu X, Yu F, Ji J. Effect of La-Fe/Si-MCM-41 catalysts and CaO additive on catalytic cracking of soybean oil for biofuel with low aromatics. J Anal Appl Pyrolysis 2019;143:104693. https://doi.org/10.1016/j.jaap.2019.104693.

[32] Gupta J, Agarwal M, Dalai AK. slurry derived hydroxyapatite as heterogeneous catalyst for biodiesel production from soybean oilMarble. Can J Chem Eng 2018;96:1873–80. https://doi.org/10.1002/cjce.23167.

[33] Likhanov VA, Lopatin OP, Yurlov AS. Study of the effective performance of the diesel engine when working on methanol and methyl ether rapeseed oil. J Phys Conf Ser 2019;1399. https://doi.org/10.1088/1742-6596/1399/5/055026.

[34] Wang B, Li S, Tian S, Feng R, Meng Y. A new solid base catalyst for the transesterification of rapeseed oil to biodiesel with methanol. Fuel 2013;104:698–703. https://doi.org/10.1016/j.fuel.2012.08.034.

[35] Georgogianni KG, Katsoulidis AK, Pomonis PJ, Manos G, Kontominas MG. Transesterification of rapeseed oil for the production of biodiesel using homogeneous and heterogeneous catalysis. Fuel Process Technol 2009;90:1016–22. https://doi.org/10.1016/j.fuproc.2009.03.002.

[36] El Khatib SA, Hanafi SA, Barakat Y, Al-Amrousi EF. Hydrotreating rice bran oil for biofuel production. Egypt J Pet 2018;27:1325–31. https://doi.org/10.1016/j.ejpe.2018.08.003.

[37] Vinoth Kanna I, Tamil Selvan R, Pinky D. An analysis of rice-bran oil as a biofuel for the four-stroke, twin-cylinder CI engine. Int J Ambient Energy 2021;42:319–24. https://doi.org/10.1080/01430750.2018.1550015.

[38] Asikin-Mijan N, Ooi JM, AbdulKareem-Alsultan G, Lee H V., Mastuli MS, Mansir N, et al. Free-H2 deoxygenation of Jatropha curcas oil into cleaner diesel-grade biofuel over coconut residue-derived activated carbon catalyst. vol. 249. Elsevier Ltd; 2020. https://doi.org/10.1016/j.jclepro.2019.119381.

[39] Olotu M. Socio-economic impact of Jatropha-based biofuel promotion on rural livelihoods in northern Tanzania 2020:1–13. https://doi.org/10.21203/rs.3.rs-25918/v1.

[40] Singh A, Sinha S, Choudhary AK, Panchal H, Elkelawy M, Sadasivuni KK. Optimization of performance and emission characteristics of CI engine fueled with Jatropha biodiesel produced using a heterogeneous catalyst (CaO). Fuel 2020;280:118611. https://doi.org/10.1016/j.fuel.2020.118611.

[41] Rodríguez-Ramírez R, Romero-Ibarra I, Vazquez-Arenas J. Synthesis of sodium zincsilicate (Na2ZnSiO4) and heterogeneous catalysis towards biodiesel production via Box-Behnken design. Fuel 2020;280:118668. https://doi.org/10.1016/j.fuel.2020.118668.

[42] Taufiq-Yap YH, Teo SH, Rashid U, Islam A, Hussien MZ, Lee KT. Transesterification of Jatropha curcas crude oil to biodiesel on calcium lanthanum mixed oxide catalyst: Effect of stoichiometric composition.

Energy Convers Manag 2014;88:1290–6. https://doi.org/10.1016/j.enconman.2013.12.075.

[43] Teo SH, Goto M, Taufiq-Yap YH. Biodiesel production from Jatropha curcas L. oil with Ca and La mixed oxide catalyst in near supercritical methanol conditions. J Supercrit Fluids 2015;104:243–50. https://doi.org/10.1016/j.supflu.2015.06.023.

[44] Patil P, Deng S, Isaac Rhodes J, Lammers PJ. Conversion of waste cooking oil to biodiesel using ferric sulfate and supercritical methanol processes. Fuel 2010;89:360–4. https://doi.org/10.1016/j.fuel.2009.05.024.

[45] Tan YH, Abdullah MO, Kansedo J, Mubarak NM, Chan YS, Nolasco-Hipolito C. Biodiesel production from used cooking oil using green solid catalyst derived from calcined fusion waste chicken and fish bones. Renew Energy 2019;139:696–706. https://doi.org/10.1016/j.renene.2019.02.110.

[46] Farooq M, Ramli A. Biodiesel production from low FFA waste cooking oil using heterogeneous catalyst derived from chicken bones. Renew Energy 2015;76:362–8. https://doi.org/10.1016/j.renene.2014.11.042.

[47] Gnanaprakasam A, Sivakumar VM, Surendhar A, Thirumarimurugan M, Kannadasan T. Recent Strategy of Biodiesel Production from Waste Cooking Oil and Process Influencing Parameters: A Review. J Energy 2013;2013:1–10. https://doi.org/10.1155/2013/926392.

[48] Ho KC, Chen CL, Hsiao PX, Wu MS, Huang CC, Chang JS. Biodiesel production from waste cooking oil by two-step catalytic conversion. Energy Procedia 2014;61:1302–5. https://doi.org/10.1016/j.egypro.2014.11.1086.

[49] Goh BHH, Ong HC, Cheah MY, Chen WH, Yu KL, Mahlia TMI. Sustainability of direct biodiesel synthesis from microalgae biomass: A critical review. Renew Sustain Energy Rev 2019;107:59–74. https://doi.org/10.1016/j.rser.2019.02.012.

[50] Damartzis T, Zabaniotou A. Thermochemical conversion of biomass to second generation biofuels through integrated process design-A review. Renew Sustain Energy Rev 2011;15:366–78. https://doi.org/10.1016/j.rser.2010.08.003.

[51] Azad KN, Salam MA, Azad KN. Aquaponics in Bangladesh: current status and future prospects. J Biosci Agric Res 2016:669–77. https://doi.org/10.18801/jbar.070216.79.

[52] Naik SN, Goud V V., Rout PK, Dalai AK. Production of first and second generation biofuels: A comprehensive review. Renew Sustain Energy Rev 2010;14:578–97. https://doi.org/10.1016/j.rser.2009.10.003.

[53] Rosen MA. Environmental sustainability tools in the biofuel industry. Biofuel Res J 2018;5:751–2. https://doi.org/10.18331/BRJ2018.5.1.2.

[54] Saravanan AP, Mathimani T, Deviram G, Rajendran K, Pugazhendhi A. Biofuel policy in India: A review of policy barriers in sustainable marketing of biofuel. J Clean Prod 2018;193:734–47. https://doi.org/10.1016/j.jclepro.2018.05.033.

[55] Brownbridge G, Azadi P, Smallbone A, Bhave A, Taylor B, Kraft M. The future viability of algae-derived biodiesel under economic and technical uncertainties. Bioresour Technol 2014;151:166–73. https://doi.org/10.1016/j.biortech.2013.10.062.

[56] Mahalingam A, Devarajan Y, Radhakrishnan S, Vellaiyan S, Nagappan B. Emissions analysis on mahua oil biodiesel and higher alcohol blends in diesel engine. Alexandria Eng J 2018;57:2627–31. https://doi.org/10.1016/j.aej.2017.07.009.

[57] Arul Gnana Dhas A, Devarajan Y, Nagappan B. Analysis of emission reduction in ethyne–biodiesel-aspirated diesel engine. Int J Green Energy 2018;15:436–40. https://doi.org/10.1080/15435075.2018.1473774.

8

A Green Automotive Industry for a Sustainable Future

Ankit Acharya and Shubhansh Jain

Undergraduate, School of Mechanical Engineering, KIIT (DU), Bhubaneswar, Orrisa, India
E-mail: acharya2000ankit@gmail.com; shubhanshjain28@gmail.com

Abstract

By the advent of the 21st century, we came across the problem of energy crisis and global warming, which made the whole scientific community rethink on the production, management, and conservation of energy and find new and sustainable sources of energy. Now a large fraction of the world's total energy produced is used by the automotive industry itself. This is so because the total world's trade is dependent on the transportation sector which is, in turn, driven by the automotive industry. So to obtain a completely green energy approach for a sustainable future it is very much necessary to make the automotive industry more energy-efficient, which will help in conserving energy and hence contribute to the dream of a sustainable future. For this purpose, in this chapter, we have included the possible enhancements in conventional Internal Combustion (IC) Engines and the futuristic technologies driven by sustainable energy sources like The Green Engine, Hybrid Electric vehicles (HEVs), and Hydrogen as a potential fuel for combustion engines (H2-ICEs). By incorporating these technologies in modern automobiles we can enhance their energy efficiency and hence can lead to a sustainable green energy-driven transportation sector in the future.

Keywords: Green engine, Hybrid vehicle, Sustainable energy, Hydrogen fuel, Green energy, Alternate fuels.

8.1 Introduction

In the last few decades, huge efforts have been made to counter the energy crisis. The most important way to do this is to regulate the production, management, and consumption of energy in every possible form. This will help us to fulfill the dream of a sustainable future. To achieve this a green energy-driven approach is needed in the development of new technologies and modification of current technologies.

Talking about the transportation sector, about 26% of total U.S. energy consumption in 2020 was for transporting people and goods from one place to another [1]. And this transportation sector depends on automobiles to transport the goods. So to achieve the objective of a sustainable future the automobiles need to be modified to be more energy-efficient, less harmful to the environment, and driven by sustainable energies. So in this chapter, we have included some of the ways to increase the energy efficiency of the currently working conventional IC Engines. And at the same time, we have included some of the emerging technologies which are driven by sustainable

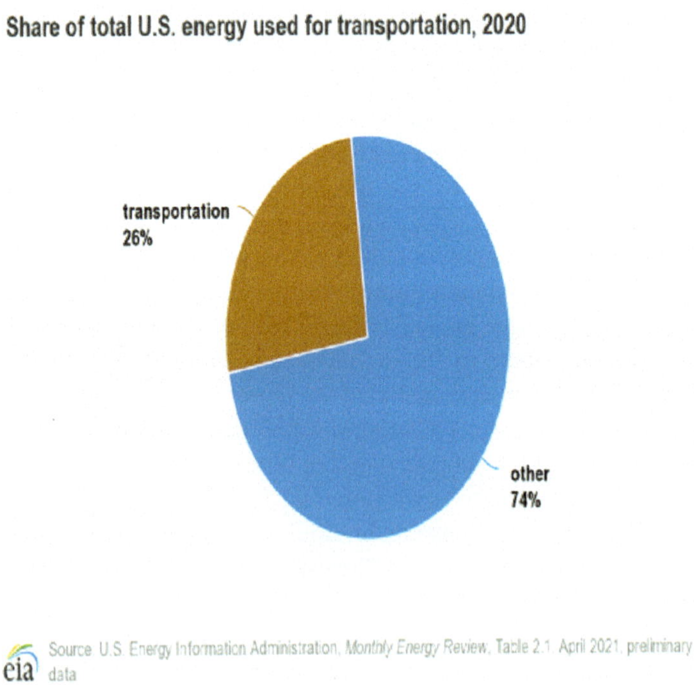

Figure 8.1 Share of total U.S energy used for transportation 2020 [1]

energies like Hybrid Electric vehicles (HEVs) and revolutionary green engine technology. With this content, we provide the modern approaches that can be inculcated into the automotive industry to enhance its energy efficiency.

Another approach to obtaining a sustainable future is to decrease the environmental pollution caused by the present technologies to the environment. Talking of the automotive industry, if the automobiles are modified to rely on energy sources that do not pose any danger to the environment then our dream of a sustainable future will certainly come true. For this, in this study, we have included the use of Hydrogen (H_2) as a potential fuel. The advantage of this would be, it would not emit any carbon dioxide (CO_2) on combustion, but will produce water (H_2O) as the by-product in its purest form. So in the next section, we are going to discuss

8.2 Scope of Development in Conventional Internal Combustion (IC) Engine

According to the International Energy Association (IEA), of the total CO_2 emissions 40% are generated during the production of electricity itself [2]. In developing countries like India and China, the major source of electricity generation is thermal energy, and it's going to continue the same for more than a couple of decades [3]. As we know, the generation of energy by burning coal in thermal power plants produces harmful products like CO,

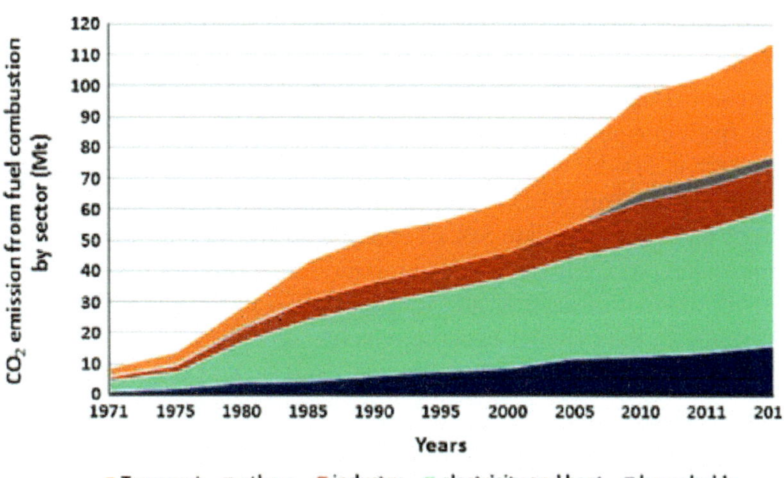

Figure 8.2 Graphical representation of CO_2 emission from 1971-2012 [1]

SO_X, and NO_X which can lead to acid rain and environmental pollution. So the switching of the whole automobile industry into Electric Vehicles (EV) will only serve the purpose of worsening the situation. The energy crisis and global warming will increase at the same time.

The only way to counter this is to develop conventional IC Engines to enhance efficiency and lower the emission characteristics. So in this section, we have included some of the possible improvements that can be incorporated on a short term and long term basis respectively.

8.2.1 Possibility of Improvement in Short Term

In current times focus of research is substantially on the enhancement of efficiency using the current market fuels available, as it is very difficult to improve both the fuels and engine technologies at the same time. Further in this section, we are going to discuss the improvements that are possible to develop the current engines by incorporating technologies that are available at this time.

8.2.1.1 Improvement in engine construction

In the short term, engines can be improved by incorporating technologies that have been proved to be efficient and cost-effective in the market. Some of the examples of such technologies are *variable geometry turbochargers, integrated exhaust manifolds, variable compression ratio, variable valve lift, cooled EGR technology, cylinder deactivation, etc.*

Going a little bit ahead of the short term (within a couple of decades), one of those technologies is the "*pre-chamber ignition system*" which is observed to give higher efficiency in reducing NO_x emission by the conventional diluting of fresh charge by the EGR system. One example of such a system is "*Mazda SKYACTIV-X*" which is an improved lean burn technology [4, 5]. SKYACTIV-X works on Spark Controlled Compression Ignition (SPCCI). When it runs on regular gasoline SPCCI compresses the fuel-air mix at a very high compression ratio, with a very lean mixture. This engine uses a spark to ignite a small amount of air-fuel mix in the cylinder this raises the pressure and temperature in the cylinder because which rest of the fuel ignites under pressure(just like in diesel engines). The *SKYACTIV-X* is a better ignition system because it can produce 30% more power and torque than the current ignition system. It has 20–30% of fuel efficiency as compared to current gasoline engines.

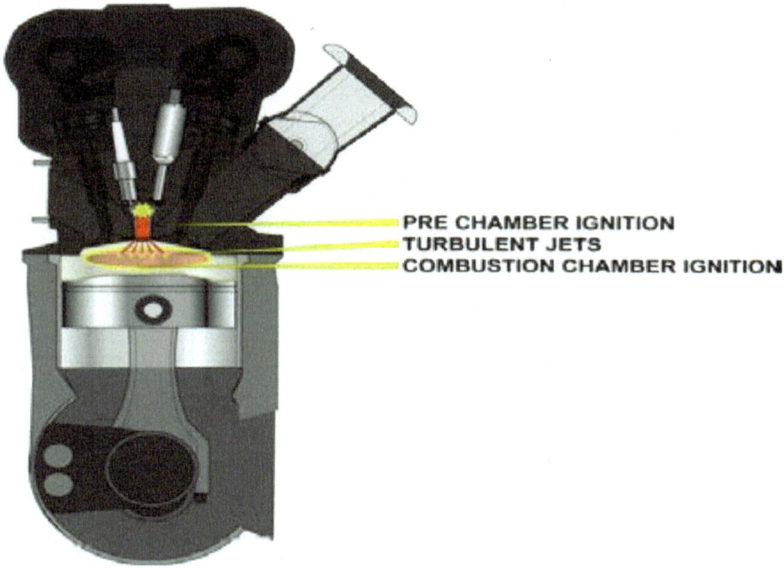

Figure 8.3 MAZDA SKYACTIV-X ignition system [45]

Similarly, another technology that can serve this purpose is the "*water injection*" -either into the intake manifold -or more effectively -directly inside the engine cylinder [6, 7]. More specifically the water takes the latent heat of vaporization from the input charge and evaporates which ensures an increase in intake charge density and lower exhaust gas temperatures. This leads to an increase in the efficiency of the engine and reduces NO_X emissions.

8.2.1.2 Exhaust treatment systems

Treatment of exhaust gases could be another way to reduce the emission of particulate matter, CO, NO_X, and hydrocarbons (HC) in the IC engines. For example, the modern "*Diesel Particulate Filter (DPFs)*" and the "*Gasoline Particulate Filters (GPFs)*" nearly separate almost all particulate particles from the exhaust of the engine [8–10]. The particulate filters trap the particulate matter such as soot and ash, they do so by using a honeycomb-like structure made up of ceramic material. The trapped impurities are then burned in fixed intervals to regenerate the filters and prevent the filter from getting choked. In certain circumstances, the DPFs even clean the exhaust gases even beyond the ambient air limits- literally cleaning the surrounding air [11].

8.2.1.3 Changes in fuel for the IC engines

As we know there are certain drawbacks of the conventional gasoline and diesel fuel for ICEs like prolonged knocking, random flame development, and propagation which leads to uncontrolled combustion. In addition to this, there are other problems like the formation of deposits inside the chamber. Also, there have been attempts to reduce the Sulphur (S_4) content in the fuel to reduce the formation of Sulphur dioxide (SO_2) in the exhaust gases, as the SO_2 is a major reason for acid rains. To counter these drawbacks the conventional fuels in IC engines can be remodified (by adding other chemicals) or replaced with synthetically developed ones to obtain higher efficiency. Some examples of them could be Ethanol, Dimethyl Ether (DME), Natural gas- in the form of both compressed (CNG) and liquefied (LNG), Ammonia (NH_3), Bio-fuel, etc. Certain additives are added to the fuels to ensure a significant reduction in the emission of sulfur oxides (SO_X-Responsible for acid rains) [12].

8.2.2 Possibility of Improvement in Long Term

There is an opportunity to develop a cheap and highly effective new engine and fuel system. Such a system can provide us with GHG benefits by fuel manufacturing using low octane products which are not processed properly. To make this kind of fuel and engine available to all, there has to be a corporation between the auto industries and oil industries and also other parties involved.

8.2.2.1 Gasoline compression ignition (GCI)

These types of ignition systems are based on the Mazda system, it does not depend on spark ignition or ignition with flame propagation. To sum up, GCI are simple diesel engines operating on gasoline-based fuels. The efficiency and exhaust emission are the same as the diesel engine but as it is less complicated so GCI's are cheaper and affordable. The fuel in GCI is less processed which makes it easier to compare to the current fuel. To make these kinds of fuels one does not need a high amount of investment from a refinery [13, 14].

The research on this engine is ongoing but it is going to take time to make it practical and for worldwide use. If used with 70 RON fuel in a GCI it lowers the carbon footprint by 5% in addition to low energy demand [15, 16]. A secondary method to use GCI is to premix air and fuel before combustion, this process might affect load and speed. These hurdles can be managed by enabling rapid combustion phasing control. There is a possibility that this engine can work with less emission and also without any kind of filter.

8.2.2.2 Reactivity controlled compression ignition (RCCI) system

It is a way to meet the requirements of different IDs under different conditions [17, 18]. A fuel with more ID like ethanol is injected with direct ignition and that activates combustion. It has a high indicated efficiency and almost zero emission of soot and NO(x) it also has a small pressure rise and noise at an acceptable level [19]. As RCCI requires two fuel injection systems it makes it expensive and complex in design. By fleet operation with centralized fuel provision, the possibility of miss-fueling is low. RCCI is most effective with commercial use, worldwide use of RCCI can help in increasing demand for diesel fuel.

8.2.2.3 Octane on demand (OOD)

The best use of available fuel octane quality is done by ODD; an engine having two injection systems will carry a low octane fuel and a high octane fuel. These components can be obtained by separately stored vehicles [20]. For getting a higher compression ratio the engine can use high octane fuel in the beginning and later a low octane fuel for the rest of the operating time.

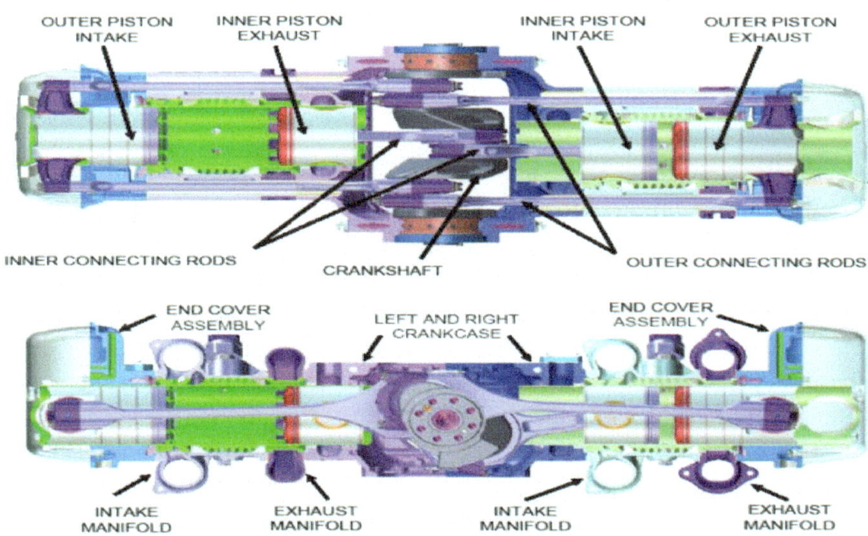

Figure 8.4 Opposed piston engines [46]

As the low octane fuel will give a lower carbon footprint which is good for the environment. The RCCI and GCI engine discussed above will also benefit from OOD as the high octane fuel will give a high load operation range and low octane fuel help in low load auto-ignition.

8.2.2.4 Opposed piston engines

They are very old engine types, they have 2 pistons in every cylinder with 1 or 2 crankshafts for each cylinder and 3 cylinders with 3 crankshafts [21]. To control the phasing to increase constant volume combustion, 2 crankshafts per cylinder are used. Low heat transfer takes place with no cylinder head which leads to high efficiency. This kind of engine operates on a weak mixture of fuel, which will provide us with the high efficiency of opposed-piston engines [22].

8.3 Green Engine Technology

In this section, we are going to explore the theory of green engine technology. The green engine is an engine concept based on a completely renewable energy source. It uses technologies like super-mixing and lean-burn approach to obtain emission characteristics hypothetically tending to zero, and efficiency is theoretically very high compared to the existing ones. Due to its very high efficiency and substantially low emissions, it is more advisable to be used in place of the conventional ICEs which have efficiencies very low (nearly 20–30%). So in the following sections, we are going to discuss the features and working of the green engine.

8.3.1 Technical features of green engine

In this section, we are going to discuss the technical features of the green engine. These features are taken into consideration during the design of the engine working process.

There are four phases in a conventional piston engine but the green engine consists of six phases, the phases are intake, compression, mixing, combustion, power, and exhaust. Some of the main features of a green engine are it has a high air change ratio, have good air to fuel mixing ratios. High combustion efficiency and many others we will discuss further along. This kind of engine is free of harmful emissions. There are no reciprocating parts so, there is no vibration.

(i) Direct Air Intake

There is no air inlet pipe, throttle, or valves in the direct air intake method. The air filter and intake port are directly connected. Hence, have high volumetric efficiency which gives high torque output.

(ii) Powerful Swirling

There is a tangential air duct between the combustion and compression chamber because this strong swirling of air is achieved. The air to fuel ratio has a combustion process satisfying working conditions.

(iii) Variable Compression Ratio

This technology is revolutionary in that it gives the best compression ratio no matter what the operating conditions are, it can work on different types of fuel and therefore have high combustion performance.

(iv) Direct Fuel Injection

It gives higher output and torque while enhancing the response for acceleration at the same time.

(v) Air-Fuel Mixing

Given enough time for mixing air and fuel under strong swirling this engine is capable of burning any liquid without any modification needed and as an ideal air-fuel mixture, it could delete CO emissions. A dense Air-Fuel mixture can help in cold start engine and manage lean burning.

(vi) High Expansion Ratio

A high expansion ratio releases much more power when fuel is burned. There is waste gas which also gives energy when burned and released. It increases engines thermal efficiency at the same time the temperature and noise decrease.

(vii) Vibration Free

There are many reciprocating parts present in the engine which produces noise and vibration. As in the green engine, most of the reciprocating parts like vanes and others are absent in it so the vibration is zero to minimal.

8.3.2 Working of Green Engine

As discussed above sections the green engine is a piston less i.e, it does not comprise any piston or any reciprocating components. It generates motive

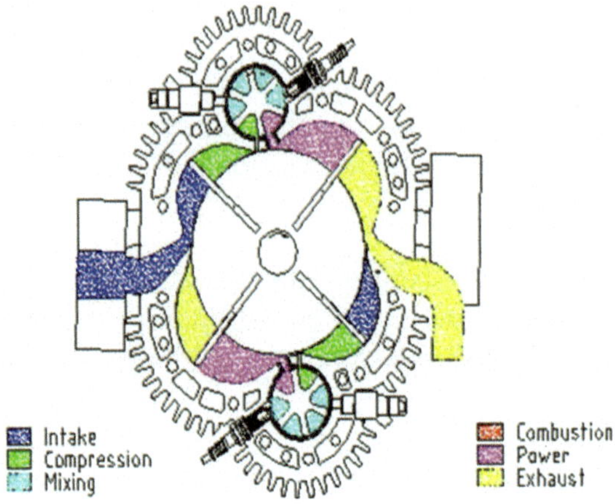

Figure 8.5 Schematic diagram of the green engine [16]

power by flowing the working fluid in a very sophisticated path in between which is combusted to increase pressure and expanded so that it can exert more pressure on the vanes in which it flows. These vanes are connected to the hub which rotates due to the pressure exerted by the fluid in the veins. It contains six phases namely intake, compression, mixing, combustion, power, and exhaust (Figure 8.5). These processes are described in detail in the following sections.

(i) Phase 1: INTAKE

The input air enters the engine directly through the air intake port, Instead of the conventional air intake system through inlet manifolds consisting of inlet valves. A small duct is provided on the sides of the vane and rotor. This duct is so designed that when air passes through it, a large swirling action is generated and the air gets compressed. This compressed air when passes through the vane blades, and the rotational moment is generated onto the blades, which are used to rotate the rotor. The intake air duct ends with a very narrow opening into the engine chamber.

(ii) Phase 2: COMPRESSION

Now the air has entered the rotor by the end of the intake process. The chambers present in the rotor are comparatively quite smaller in volume, so the air further gets compressed. The compression ratio obtained by this process is sufficient enough. Note that the air is still in swirling motion inside the rotor.

(iii) Phase 3: MIXING

In the rotor chamber, when the air comes in front of the fuel injector, the system is timed to spray the fuel onto the compressed air, at that particular instant of time. Because of the unique design of the chamber and the turbulent flow of the compressed air generated by passing through the inlet duct and chambers of the rotor, the air and fuel are mixed well. The Spark plug is located comparatively at a far distance, which provides sufficient time for the air and fuel to mix properly. This complete mixing of air and fuel fulfills the super-mixing feature of the green engine. This also minimizes the possibility of the production of unburnt carbon particulates and carbon monoxide (CO) as combustion by-products.

(iv) Phase 4: COMBUSTION

By the end of the mixing phase, the air-fuel mixture and the rotor reach the spark plug. The spark plug then at that particular instant generates a spark that combusts the air-fuel mixture. Due to the super-mixing of air and fuel, lean combustion is obtained, which is necessary to reduce the chance of any unburnt fuel particles being present in the exhaust gases. After the charge gets ignited, a uniform flame-front is generated, ensuring controlled combustion. As soon as the whole charge gets combusted, the rotor aligns itself back to the exit of the air duct, from where the entire process of mixing and combustion can be repeated.

(v) Phase 5: POWER

After the combustion process is completed, the working fluid now has immense pressure generated within it. This fluid is then allowed to pass through a small duct directing to the expansion chamber. The volume of the expansion chamber is comparatively very high. So when the fluid exits the duct to enter the expansion chamber, it experiences a sudden increase in volume. This allows the expansion of the gases to the maximum limit, ensuring nearly complete use of the thermal energy. Now when this gas expands in the vanes of the turbine, it generates sufficient force on the main turbine to rotate and generate the motive power, which is used to rotate the wheels and to do other auxiliary engine operations like running the rotor in the rotor chamber.

(vi) Phase 6: EXHAUST

As the complete thermal energy has been utilized, the exhaust gases carry with them very little or nearly zero thermal energy. This helps in increasing the thermal efficiency of the engine as the loss of thermal energy is minimized. Also as lean combustion is obtained, the temperature inside the combustion chamber does not reach much higher values. So the production of NO_X is not possible. So the lean-burn ensures that poisonous gases like CO and NO_X are absent in the exhaust gases.

8.4 Hybrid Vehicles (HVs)

We know that Conventional IC engines have been existing since the 19th century itself and so is also true for all-electric vehicles, as the first electric vehicle technology emerged towards the end of the 19th century. A lot of enhancements have also happened in these technologies separately since their emergence. And in the current 21st century both of them have emerged as the prime source of motive power for automobiles and will continue to do so for the next few years. But to achieve our goal of a sustainable future, we need more sophisticated and efficient technology. This eventually led to the discovery of the technology of the Hybrid Vehicle, which is a combination of both the conventional parent technologies. So in this section, we are going to discuss in detail the technology of Hybrid vehicles, their types, and certain pros and cons of it.

8.4.1 The Definition of Hybrid Vehicles (HVs)

A hybrid vehicle [23] uses two or more different types of propulsion systems to propel itself. For example, a vehicle can be propelled by using two propulsion systems, the combination of the electric drivetrain and the conventional

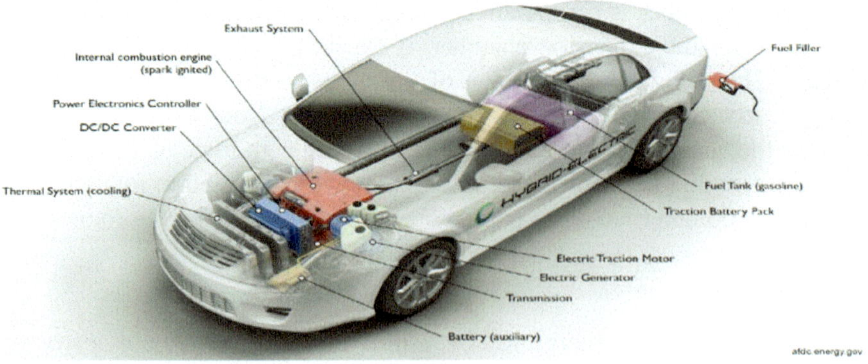

Figure 8.6 Schematic diagram of hybrid vehicle [24]

engine drivetrain. Another example could be the submarines, they use battery propulsion when submerged and diesel propulsion when surfaced.

The basic concept behind hybrid technology is that the two propulsion systems are themselves very efficient in separate conditions. Like the electric motor is very efficient in producing instant torque and power, which is required during starting. And the engine's propulsion is very good at maintaining high speeds for a long time. So if the propulsion systems are switched at the proper time, the efficiency of the vehicle can be enhanced to a very high value.

The batteries in the HVs are charged by different sources like directly from the IC engine itself or, by a Plug-in system, from solar energy, or by the Brake Energy Regeneration process. Based on these Charging methods of batteries the HVs are classified into three types, which are: Hybrid Electric Vehicle (HEVs), Hybrid Solar Vehicles (HSVs), and Plug-in Hybrid Electric Vehicles (PHEVs). Each one of these three types of HVs is discussed in further sections in detail.

8.4.2 Types of Hybrid Vehicles

8.4.2.1 Hybrid electric vehicles (HEVs)

It comprises both the IC Engine propulsion and electric propulsion systems in it. For that, the vehicle contains a fuel tank with conventional fuels to run the IC Engine and along with it has a battery that drives the electric motor for the electric propulsion system. The batteries in this case are charged either only by the IC engine or by both Engine and *Brake Energy Regeneration(BER) process*. The mechanical and battery propulsion systems are connected in four different ways to obtain four different unique drive trains [23], which are the Series hybrid, Parallel hybrid, and Series-Parallel hybrid respectively(refer to the following fig. for layout). We are going to study the layout of each one of them in the further sub-sections

(i) Series Hybrid Electric Drivetrain

A series hybrid device has two power sources feeding a single power plant that propels the automobile. It has a unidirectional energy source and also a unidirectional energy converter. This device can be used on any kind of speed and torque. With the help of optimum design and simulation, the efficiency is improved and harmful emissions are reduced. One of the major problems with this device is that the energy from the engine is converted twice due to which energy loss is high and the generator makes the engine heavier. In this

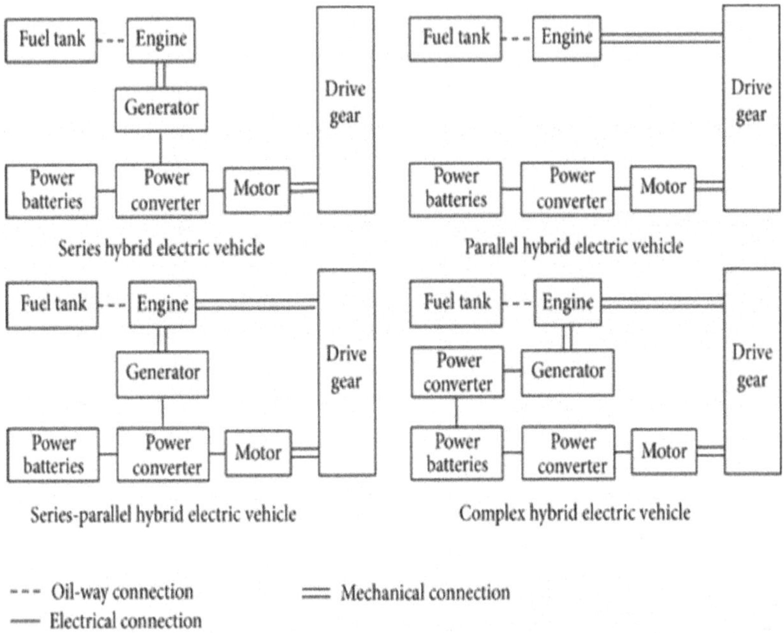

Figure 8.7 Line diagram of different HV powertrains [26]

drivetrain, the Engine generates power which charges the batteries which then powers the motor which drives the gear train. The engine does not directly supply power to the wheels. For example- *Toyota Prius*, has a total range of 640 miles on a single charge.

(ii) Parallel Hybrid Electric Drivetrain

This device supplies energy mechanically to the wheel like IC engine vehicles. This device is accompanied by an electric motor and is coupled together by mechanical coupling. Due to this configuration and combination, there is enough room for modifications in this design. There are two ways to operate this drive train, one is that the wheels are powered by the fuel tank and the engine the other one is gear train uses power batteries, and a motor which can power the gear train. Example-*BMW i-8*. It has an impressive mileage and can save upto 4700$.

(iii) Series-Parallel Hybrid Electric Drivetrain

This is a layout when both parallel and series hybrid devices are merged in such a way that the device uses the advantages of both the dives at the same time. This will allow for high efficiency in the device and have low emission from the exhaust. It's similar to the series hybrid electric drive train with one simple change, i.e one can shift from electric to fuel use using the drivetrain as per their requirement. Example-*HONDA INSIGHT, TOYOTA PIRUS*. It has a range of 150 miles.

(iv)Complex Hybrid Electric Drivetrain

An example of this device is when the front wheels are powered by hybrid propulsion while the real wheels are powered by a purely electrical system. It has 2 separate mechanical links giving flexible mounting and light transmission. In this the engine generates power then with the help of the power converter it gets stored in the battery which is used in the motor but before entering the motor the power gets converted again into usable power which then further is used in the drivetrain. Example-*Tesla Model S (Plaid)*. It can go up to 65 miles in 1.99 seconds.

8.4.2.2 Hybrid solar vehicle (HSVs)

This is another kind of hybrid vehicle that is obtained by the combination of the conventional IC engine propulsion system and the solar energy propulsion system generated by the use of photovoltaic cells. The photovoltaic cells are normally placed on the roof of the vehicle in the form of solar panels [23].

These HVs are also characterized as the connection of the two propulsion systems, which are Series, Parallel, Series-Parallel, and Complex Hybrids

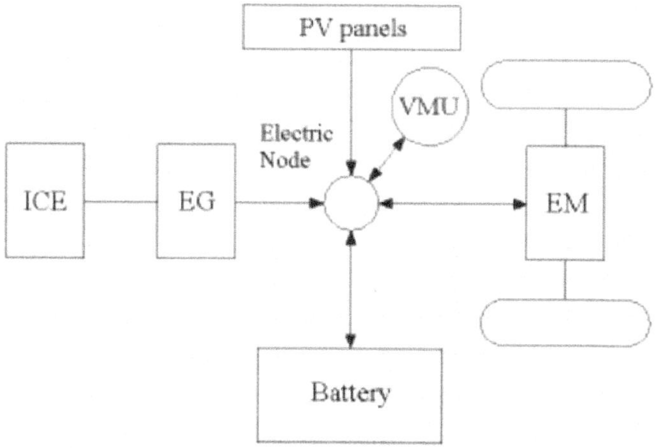

Figure 8.8 Line diagram of the Hybrid Solar Vehicle [23]

respectively. Of these four categories of HVs, the series Hybrid type is the most efficient of its contemporaries. The figure shows the layout sketch of the Series Solar Hybrid Vehicle. This type of drivetrain has been seen in cars like *STELLA LUX* and *STELLA ERA* models. They proved their capability in the market by producing a range of 1500kms. in a single charge.

8.4.2.3 Plug-in-hybrid electric vehicle (PHEVs)

This is yet another type of Hybrid vehicle, and this is the most popular one in the market at the moment. It's a Hybrid vehicle in which we use a larger battery i.e, the use of a battery propulsion system is higher compared to the IC Engine part. Here the battery is not only charged by the IC engine Brake energy regeneration but, along with it, a plug-in option is also provided. This can be plugged into any 110V/ 230V outlet for charging the battery. Most of today's HVs use this technology as it is much more efficient and cleaner than its contemporaries.

This type of powertrain is the most prevalent in current times, for example, the *TESLA MODEL S* is one of the best EVs being produced at the current time. Its Long Range model is claimed to have an impressive range of 620 miles and an impressive (0-60) miles/s in only 1.99 seconds.

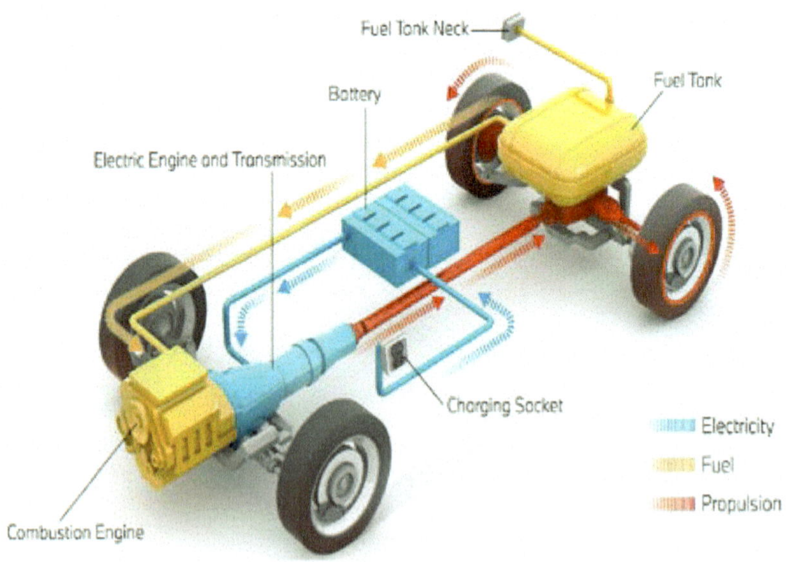

Figure 8.9 Schematic representation of PHEC's [43]

The plug-in HVs can be characterized as follows-

(a) Series Plug-in Hybrids

These HVs are also termed *Extended Range Electric Vehicles* (EREVs). These HVs use only the batteries to generate the motive power, and the engine is only used to charge the battery only when the battery dries up during running. And during braking also the battery is charged by the BER process. The battery can also be charged to full by its plug-in option. If only shorter trips are desired, then there could be no need to operate the engine, as the car can run on the battery itself.

(b)Parallel Plug-in Hybrids

These hybrids are also termed Blended Plug-in Hybrids because in these both the engine propulsion and battery propulsion are mechanically connected to the wheels. i.e It is driven by both the battery and the engine. Now the battery is charged by the IC engine during free motion. And During braking, the batteries are also charged by the BER process. As it is a PHEV, a plug-in option is also available to charge the battery.

8.4.3 Need HVs to Replace Conventional ICs and EVs-Why & Why Not??

The electric engine turns itself off during idle cycle instead it uses a gas motor at low speed and they perform well at high speed too, the gas motor can give more power for a given motor weight. The added benefit of using an electric motor is that it gives less emission in rush hour traffic. When this automobile is working on a gas motor at a time the battery can charge back up. An electric car owner can get stranded on the side of the road because his car run out of charge, but this is not the case for the hybrid car owners as their car runs out of charge the gas motor will take its place automatically. HVs are the most gas-efficient cars ever to be produced. Over a decade hybrid vehicles can save upto $2,500 as compared to gasoline vehicles. Sure, some of the reasons for increased efficiency are aerodynamics, the weight of the engine, and better mileage by reducing engine size. In hybrid manufacturing, the Japanese are the recognized leaders. The most selling hybrid cars are *Honda Insight and Toyota Prius*.

Now, let us talk about why HVs should not be bought. As discussed above the engine is very heavy and complex which means the failure will be present and as the device is complex it will take more time to render the failure and it will also take a huge amount of money to render it. These failures will occur less as the automobile is connected to a computer that takes care of everything. The main reason for a customer not buying a hybrid is that they want power and zippiness in their car like Dodge Demon or other American muscle cars and sports cars. Secondly, the mileage of these cars is less than advertised. Some of the drivers noticed that the hybrid vehicles have 10% less mileage than advertised in the market. That is one of the reasons people don't buy these kinds of automobiles.

8.5 Hydrogen Fuel IC Engines (H_2-ICEs)

In this section, we are going to discuss the technology of H_2-ICEs. For this, we use Hydrogen (H_2) as a fuel in the conventional IC engines. Now, why only H_2? The answer to this question lies in the chemical properties of hydrogen. As we know, combustion is an oxidation process, and when hydrogen goes through oxidation it results in pure water (eq. 8.1). So by the use of hydrogen, we not only save energy but also produce water (in its purest form), which is the true sense of a sustainable energy approach. Also, the combustion of H_2 is very lean, resulting in the decrease of production of $NO_{(X)}$ to minimum levels. So the use of hydrogen as a fuel in IC engines is the best way to make the automotive sector sustainable.

$$\{2H_2(g) + O_2(g) \rightarrow 2H_2O(g)\} \tag{8.1}$$

8.5.1 Fundamental of H_2-ICEs

The major function of this engine is to burn cleanly and operate efficiently, this is due to the unique characteristics of hydrogen which can give ultra-lean combustion and can drastically reduce NO_x production. It does impose some technical challenges that are at high engine loads due to the propensity of pre ignite the hydrogen-air mixture and increased NO_x production. The properties of this engine can be compared with the PFI gasoline engine. We will talk more about this technology in the below-mentioned topics. It works with the hydrogen extracted from water molecules by removing Oxygen and storing Hydrogen. We can use different sources for Hydrogen but water is the most convenient one. To run this kind of engine one need platinum rods and other components made of platinum which is expensive. When the vehicle is running it will give a low amount of NO_x emission, but while making H-based fuel the industries will devour a lot of energy. Also, the storage of hydrogen is a matter of concern, as the H_2 has to be stored only in liquid form, and this will require special pressure vessels which will need to be stored in very low temperatures. This will again cost a lot more energy to the vehicle for the refrigeration purpose of the fuel tanks. This kind of makes the point of Hydrogen based engines moot.

8.5.2 Types of Advanced H_2-ICEs

Higher power density can be achieved with PFI-H_2ICE *(PFI-Ported Fuel Injection)* premixed naturally. In practical use, the power density loss can be as high as 50% [25]. In this section, we will talk about how the power and efficiency of H-type IC engines can be increased and $NO_{(x)}$ emissions can be decreased. As discussed above we need platinum to make some components Researches are trying very hard to find a replacement for Platinum. At MIT they have found a Replacement for platinum which is *Cobalt Phosphide* [26]. A lot of research is going on which will be updated soon and a new variant of H based engine will be available. If we use *Cobalt Phosphide* instead of H-based it will give a high amount of energy and it is cheap to produce as its raw materials are present in abundance in nature.

8.5.2.1 Pressure based H_2ICE

Boosting pressure while air intake is a great way to increase the peak engine power in conventional IC engines. For premixed or PFI-H2ICEs, boosting pressure is important to achieve higher power than conventional IC engines because of increasing pressure we can achieve higher power generation in the engine. Which will give us higher speed and greater efficiency. Due to low pressure it will give less emission and give in its hydrogen-based IC engine its by-product will be water. Different studies were going on this topic by many different authors [27–28]. Their research was the basis of today's turbocharged engines and their work is still being continued in the hope of developing sustainable H_2ICEs.

The most recent breakthrough was done by *BMW and Ford* [29, 30]. According to the reports from *BMW*, a single-cylinder engine supercharged to 1.8 bar achieves a 30% increase in specific power output compared to a naturally aspirated gasoline engine. From another report from *FORD ZETEX ENGINE* and a *4-cylinder2.3-FORD DURATEC ENGINE* that is used for conventional and hybrid vehicles. At *Mitsubishi Institute of Technology*, 2 *Nissan* engines were tested for Hydrogen-Hybrid Engine, which showed a similar 35% increase in power due to boosting while holding NO_x emissions at 10 ppm [31].

8.5.2.2 Liquid-hydrogen-fueled internal combustion engine (l-H_2-ICEs)

The use of liquid hydrogen has many benefits over the gaseous fuel port injection in conventional H_2ICEs [32]. In this process, the liquified H_2 fuel is not directly supplied to the engine but only stored in liquid form for easy storage. The use of liquefied fuel ensures the storage of very high energy density as compared to the gaseous ones. And this also helps in decreasing the temperature of the combustion chamber which leads to a lesser chance of $NO_{(X)}$ production.

The use of liquid H_2 ensures *Intake-charge cooling,* which improves the volumetric efficiency (Since a higher amount of fuel is present in the same volume). This also helps in minimizing the pre-ignition of charge preventing knocking. Now as the volumetric efficiency increases, the output power density also increases subsequently. A substantial amount of research work is out there which proves this, for example, it has been calculated that liquid hydrogen at a temperature of 120 K when used, the peak power output of the l-H_2ICEs can equal that of the identical engine(in terms of size parameters like the compression ratio, etc) fueled with gasoline [33]. Similarly, another study suggests that with intake charge cooling to 210 K, the power density of an L-H2ICE will be 15% higher compared to fueling with a conventional PFI gasoline engine [34].

8.5.2.3 Direct-injection hydrogen-fueled internal combustion engine (DI-H2ICE)

It has been viewed as one of the most advanced H_2-based IC engine options [35]. This view is based on volumetric efficiency and a great potential to avoid pre-ignition. It was measured that there is an increase of 15% in IMEP when using H based engine instead of an IC engine. The main objective of DI-H_2ICE is that it requires the mixing of Air and Hydrogen in a short period. the plausibility of complete mixing in a DI-H_2ICE, Homan [36], by using experimental relations for air trapping rates in free turbulent jets so that the magnitude expressions for turbulent mixing times, let us assume that a free-hydrogen jet with the sonic velocity at the orifice issuing into the air will entrain a stoichiometric amount of air in approximately 1ms. However, contrary to this optimistic estimate, the overwhelming experimental evidence [37, 38–41] demonstrates that complete mixing in an engine takes approximately 10ms. In the case of late injection, there is a probability of incomplete mixing with late injection, much effort has been devoted to understanding the effect of injection timing on DI-H_2ICE properties. The effect of NO_x emissions is reduced.

8.5.2.4 H_2-ICE-electric hybrid

The electric version of H_2ICE offers improved efficiency and low emission without any need for after treatment. Recent development in the Hydrogen-Electric hybrid shows that it can work on both series and parallel configurations. This configuration is used to drive an alternator that generates electricity, this electricity is used to power the batteries or can be used for driving the motor which powers the drive train. Modeled off a non-hybrid H_2ICE vehicle, and parallel and series H_2ICE HEVs, they reported that for the input parameters selected, all three vehicles can satisfy emissions standards. An attractive feature of the HEV is that the peak power output of the ICE can be significantly lower than that required for a non-hybrid ICE without any sacrifice of vehicle performance [42].

One of the major advancements will be the *Ford* which is using the H_2RV engine which was modified as an HEV type engine that uses a supercharged 4-cylinder H_2ICE. The boosted H2ICE has a peak power of 110 HP at 4500 rpm and the electric motor provides an additional 33 HP and is used primarily for power assist. Acceleration is 0–60 mph in 11 s. [44] which is quite impressive for an HEV.

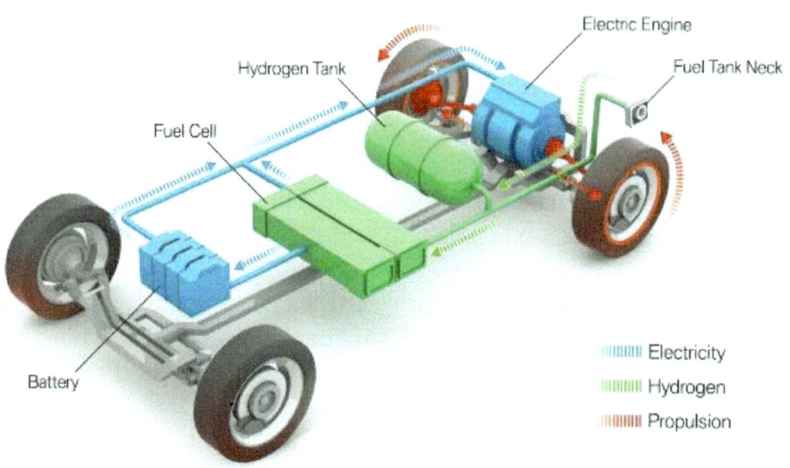

Figure 8.10 Diagram of H_2ICE electric hybrid powertrain [28]

8.6 Conclusion

Over the last couple of decades, there has been a huge amount of study done in the arena of sustainable energy. These sustainable sources of energy have been present with us for a long time, but they have started to be taken seriously in the last couple of decades only. Now to obtain a completely sustainable future it is also very important to modify the prevailing technologies to be greener and energy-efficient, along with the adoption of new sustainable sources of energy as discussed in various sections in this chapter.

So to make the automotive industry more sustainable in any country, we need to assess the sources of energy generation in it. For example, if we take the case of developing countries like India, here we are largely dependent on thermal power plants for electricity generation. So if the automobile industry is made to go completely electric, then it will just increase the load on thermal power plants for electricity generation, which will lead to more pollution. Also, we lack the required infrastructure for the completely electricity-based automotive industry and the use of hydrogen vehicles cannot be commercialized due to its storage problems. So we need to continue with the traditional ICEs for the propulsion of automobiles and with that, we should focus on decreasing the dependency on thermal power plants for electricity generation. Also, we should focus on developing the infrastructure that is required for the modern EVs to operate efficiently i.e increase the facility of charging centers all over the country. Now as the ICEs are the prime mover in the developing countries, they are needed to be more energy-efficient and less polluting to save energy for a sustainable future (as discussed in sec.2).

Now taking developed countries like the US and Russia into consideration, they mostly use sustainable sources of energy for electricity generation. So they can go fully electric in the automotive industry and can phase out the conventional ICEs completely, except for those which are unavoidable like marine and military applications. Also, hybrid powertrains can be used (as discussed in sec.4).

Now to make the whole world's automotive industry to be green energy driven, the suggestions given in previous paragraphs can be taken into consideration. And along with that, the scientific community should try to make the H2 based ICEs and hydrogen fuel cell technologies more efficient by eliminating the major drawback of fuel storage. Along with that, Green engine technology can be developed to be commercialized in the future for the automotive industry to be more sustainable. By adopting the given technologies in this study, we can surely make the automotive industry to be based on green energy for a sustainable future.

Acknowledgement

We sincerely pay our gratitude to Dr. A. K. Rout, Associate Professor, KIIT Deemed University, for advising and guiding us throughout this work. His enthusiasm, knowledge, and exacting attention to detail have been an inspiration and kept our work on track. We would like to pay our gratitude to the editors of this book for providing us with this golden opportunity to contribute something to this book.

References

[1] US Energy Information Administration website

[2] International Energy Agency (IEA), CO2 emissions from fuel combustion – Overview, 2018,Available-from: https://www.iea.org/newsroom/events/statistics--co2-emissions-from-fuelcombustion-overview-.html

[3] Kalghatgi, G.T., Is it really the end of internal combustion engines and petroleum in transport?AppliedEnergy, 2018. 225: p. 965-974, doi:10.1016/j.apenergy.2018.05.076

[4] Bunce M, et al., Development of a Light Duty Gasoline Engine Incorporating Jet Ignition for Stable Ultra-Lean Operation. JSAE Paper 20195413, 2019.

[5] Sens, M., et al. Pre-Chamber Ignition and Promising Complementary Technologies. in 27th Aachen Colloquium, Aachen. 2018

[6] De Cesare, M., et al., Technology Comparison for Spark Ignition Engines of New Generation. SAE Int. J. Engines 10(5):2513-2534, 2017, doi:10.4271/2017-24-0151

[7] Hoppe, F., et al., Evaluation of the Potential of Water Injection for Gasoline Engines, in SAE International Journal of Engines. SAE Int. J. Engines 10(5):2500-2512, 2017, doi:10.4271/2017-24-0149

[8] Johnson, T. and A. Joshi, Review of Vehicle Engine Efficiency and Emissions. SAE Int. J. Engines 11(6):1307-1330, 2018, doi:10.4271/2018-01-0329

[9] ACEA. Diesel: new data proves that modern diesel cars emit low pollutant emissions ontheroad,2017.Availablefrom: https://www.acea.be/press-releases/article/diesel-nw-dataproves-that-modern-diesel-cars-emit-low-pollutant-emissions (accessed: 24 April 2019)

[10] AECC. Gasoline particulate filter (GPF) - How can the GPF cut emissions of ultrafine particles

[11] fromgasolineengines?2017.Availablefrom: https://www.aecc.eu/wpcontent/uploads/2017/11/2017-AECC-technical-summary-on-GPF-final.pdf (accessed: 14May 2020)

[12] Kalghatgi, G.T. and R. Stone, Fuel requirements of spark ignition engines. Proceedings of the Institution of Mechanical Engineers, Part D: Journal of Automobile Engineering, 2017. 232(1): p. 22-35. doi:10.1177/0954407016684741

[13] Technical Committee of the Petroleum Additive Manufacturers in Europe (ATC), Fuel Additives: Uses and Benefits. 2013. Available from: https://www.atceurope.org/public/Doc113%202013-11-20.pdf

[14] Kalghatgi, G., et al., The outlook for transport fuels: Part 1, in Petroleum Technology Quarterly (PTQ). 2016. p. 23-31.

[15] alghatgi, G., et al., The outlook for transport fuels: Part 2, in Petroleum Technology Quarterly (PTQ). 2016. p. 17-23.

[16] Lu, Z., et al., Well-to-Wheels Analysis of the Greenhouse Gas Emissions and Energy Use of Vehicles with Gasoline Compression Ignition Engines on Low Octane Gasoline-Like Fuel. SAE Int. J. Fuels Lubr. 9(3):527-545, 2016, doi:10.4271/2016-01-2208

[17] Hao, H., et al., Compression ignition of low-octane gasoline: Life cycle energy consumption and greenhouse gas emissions. Applied Energy, 2016. 181: p. 391-398, doi: 10.1016/j.apenergy.2016.08.100

[18] Splitter, D., et al., Reactivity Controlled Compression Ignition (RCCI) Heavy-Duty Engine Operation at Mid-and High-Loads with Conventional and Alternative Fuels. SAE Technical Paper 2011-01-0363, 2011, doi:10.4271/2011-01-0363

[19] Kokjohn, S.L., et al., Fuel reactivity controlled compression ignition (RCCI): a pathway to controlled high-efficiency clean combustion. International Journal of Engine Research, 2011. 12(3): p. 209-226, doi: 10.1177/1468087411401548

[20] Nieman, D.E., et al., Heavy-Duty RCCI Operation Using Natural Gas and Diesel. SAE Int. J. Engines 5(2):270-285, 2012, doi:10.4271/2012-01-0379

[21] Chang, J., et al., Octane-on-Demand as an Enabler for Highly Efficient Spark Ignition Engines and Greenhouse Gas Emissions Improvement. SAE Technical Paper 2015-01-1264, 2015, doi:10.4271/2015-01-1264

[22] Chatterton, E., The Napier Deltic Diesel Engine. SAE Technical Paper 560038, 1956, doi:10.4271/560038

[23] Hybrid Vehicle: A Study on TechnologyKaran C. Prajapati 1,*, Ravi Patel 2 and Rachit Sagar 3 1, ,3: B.Tech (Mechanical Engineering)

Student,Department of Mechanical Engineering,School of Technology, Pandit Deendayal Petroleum University Raisin, Gandhinagar, Gujarat - 382007, India

[24] The hydrogen-fueled internal combustion engine: a technical review C.M. White∗, R.R. Steeper, A.E. Lutz Sandia National Laboratories, Combustion Research Facility, P.O. Box 969, MS 9053, Livermore, CA 94551-0969, USA

[25] Stockhausen WF, Natkin RJ, Kabat DM, Reams L, Tang X, Hashemi S, et al. Ford P2000 hydrogen engine design and vehicle development program. SAE paper 2002; 2002-01-0240.

[26] Toward High-Performance and Low-Cost Hydrogen Evolution Reaction Electrocatalysts: Nanostructuring Cobalt Phosphide (CoP) Particles on Carbon Fiber Paper Shu Hearn Yu and Daniel H. C. Chua*

[27] Nagalingam B, Dübel M, Schmillen K. Performance of the supercharged spark ignition hydrogen engine. SAE paper 1983; 831688.

[28] Furuhama S, Fukuma T. High output power hydrogen engine with high pressure fuel injection, hot surface ignition and turbocharging. Int J Hydrogen Energy 1986;11:399–407.

[29] Natkin RJ, Tang X, Boyer B, Oltmans B, Denlinger A, Heffel JW. Hydrogen IC engine boosting performance and NO*x* study. SAE paper 2003; 2003-01-0631.

[30] Jaura AK, Ortmann W, Stuntz R, Natkin B, Grabowski T. Ford's H2RV: an industry first HEV propelled with an H2 fueled engine—a fuel efficient and clean solution for sustainable mobility. SAE paper 2004; 2004-01-0058.

[31] A technical survey of contemporary U.S. projects. Technical Report, Escher Technology Associates, Inc., Report for the US. Energy and Development Administration, Report No. TEC74/005, 1975.

[32] Peshka W. Hydrogen: the future cryofuel in internal combustion engines. Int J Hydrogen Energy 1998;23:27–43.

[33] Furuhama S, Hiruma M, Enomoto Y. Development of a liquid hydrogen car. Int J Hydrogen Energy 1978;3:61–81.

[34] Wallner T, Wimmer A, Gerbig F, Fickel HC. The hydrogen combustion engine: a basic concept study. In: Gasfahrzeuge Die passende Antwort auf die CO2-Herausforderung der Zukunft?, Expert-Verlag GmbH, 2004.

[35] Homan HS. An experimental study of reciprocating internal combustion engines operated on hydrogen. Ph.D. Thesis, Cornell University, Engineering, Mechanical; 1978

[36] Homan HS. An experimental study of reciprocating internal combustion engines operated on hydrogen. Ph.D. Thesis, Cornell University, Engineering, Mechanical; 1978

[37] Swain MR, Schade GJ, Swain MN. Design and testing of a dedicated hydrogen-fueled engine. SAE paper 1996; 961077.

[38] Homan HS. An experimental study of reciprocating internal combustion engines operated on hydrogen. Ph.D. Thesis, Cornell University, Engineering, Mechanical; 1978

[39] Glasson ND, Green RK. Performance of a spark-ignition engine fuelled with hydrogen using a high-pressure injector. Int J Hydrogen Energy 1994;19:917–23.

[40] Jorach R, Enderle C, Decker R. Development of a low-NO*x* truck hydrogen engine with high specific power output. Int J Hydrogen Energy 1997;22:423–7.

[41] Kim JM, Kim YT, Lee JT, Lee SY. Performance characteristics of hydrogen fueled engine with the direct injection and spark ignition system, SAE paper 1995; 952498.

[42] Aceves SM, Smith JR. Hybrid and conventional hydrogen engine vehicles that meet EZEV emissions. SAE paper 1997; 970290.

[43] Keller J, Lutz A. Hydrogen fueled engines in hybrid vehicles. SAE paper 2001; 2001-01-0546.

[44] Natkin RJ, Tang X, Boyer B, Oltmans B, Denlinger A, Heffel JW. Hydrogen IC engine boosting performance and NO*x* study. SAE paper 2003; 2003-01-0631.

[45] https://drivetribe.com/p/f1-burning-lean-pre-chamber-combustion-VMyP7dZLRZmQwvUma3ogBQ?hcb=1&iid=J-CskszvSM6N6k6Mhtzpbg

[46] https://www.greencarcongress.com/2005/05/fev_developing_.html?hcb=1

9

Thermochemical Conversions of Contaminated Biomass for Sustainable Phytoremediation

Khanh-Quang Tran[1,*] and Zhongchuang Liu[2]

[1]Department of Energy and Process Engineering,
Norwegian University of Science and Technology, Kolbjørn Hejes vei 1B, Trondheim, Norway
[2]Green Intelligence Environmental School, Yangtze Normal University, 16 Juxian Rd. Lidu, Fuling District of Chongqing, China
E-mail: khanh-quang.tran@ntnu.no
*Corresponding Author

Abstract

Phytoremediation of HM-contaminated soils yields plant biomass, which should be safely disposed of. This chapter provides an overview and references for technical parameters and characteristics of a variety of disposals and their application strategies for phytoremediation plants contaminated with HMs using thermochemical conversion processes such as combustion, gasification, and pyrolysis. Through combustion, gasification, or pyrolysis, plant biomass can be reduced with HMs. The techniques for obtaining solid, liquid, and gas by-products can be recycled using various methods. Each dumping and utilization option has its own set of advantages and disadvantages. Combining several strategies may allow you to maximize their benefits while minimizing their drawbacks, which could be a realistic alternative.

Keywords: Biomass, Combustion, Gasification, Emissions.

9.1 Introduction

Recent decades have seen increased interest in biomass as a fuel for heat and electricity generation due to rising worries about future energy supplies and the need to decrease CO2 emissions. The increasing burning of plant biomass materials to produce heat and electricity has prompted the restoration of mineral nutrients contained in biomass ash to forest and agricultural soils. This is partly because soil chemistry estimates indicate that ground weathering may not always be capable of releasing mineral nutrients at the same rate as chemicals extracted from the forest during biomass production and harvesting. This requirement is essential for calcium, magnesium, and potassium base cations. Long-term, there is a concern that the soils' ability to resist acidification will be harmed due to air emissions of acidic gases from combustion plants and a lack of nutrients essential to plants. As a result, it's critical to return nutrients to the soils by spreading biomass ash from the combustion process onto the ground. Except for nitrogen, the ash retains all of the minerals and nutrients previously present in the biomass. The mineral losses that occurred when the biomass was removed from the forest are compensated when the ash is restored to the forest. Because ash is basic, it helps to prevent soil acidification. The use of biomass as a fuel for heat and electricity generation will be sustainable only when biomass ash is returned to the soils in an amount corresponding to the biomass harvested. Due to large quantities of dangerous heavy metals in some ash fractions, which may pollute soils, recycling the entire amount of biomass ash produced from combustion is problematic.

On the other hand, soil pollution is one of the world's most critical environmental problems, causing serious health and environmental risks [1–3]. For example, over 3 million soil sites are suspected of being contaminated in the European Union, with 250,000 requiring treatment [4]. Heavy metal-contaminated soils, such as Cd, Cr, Pb, Cu, Zn, Co, Ni, Se, Cs, and As, account for more than 37% of instances, followed by mineral oil contamination (33.7%), polycyclic aromatic hydrocarbon contamination (13.3%), and other pollutants [5]. This might have major ramifications for the global economy and geopolitics in the future years because of the negative effects of heavy metals on human and environmental health when they are introduced into the food chain or by dust inhalation and swallowing of contaminated soil particles [3, 6].

Soil remediation for heavy metal contamination provides unique scientific and technological obstacles since heavy metals cannot be further converted into non-harmful molecules. As a consequence, the only way to clean up the soil is to remove or sequester heavy metals. Chemical treatment in-situ or ex-situ, bioremediation, chemical methods, soil flushing, vitrification, incineration, and landfilling are only a few of the current technical solutions for heavy metals cleanup [7].

Extraction of pollutants from polluted regions, the transformation of metal elements into less dangerous chemical species, or sequestration of metal elements in roots to avoid leaching are all examples of in-situ biological treatment approaches (phytostabilisation). Phytoremediation is an exciting new technique that may be used in place of more expensive physical and chemical treatment approaches [8]. Phytoremediation is also an environmentally friendly technology because it removes toxins while conserving plant metabolism processes without changing the soil's physical, chemical, or ecological characteristics [9]. On the other hand, Phytoremediation is limited as a biological approach by various factors, including the lengthy treatment time, the location, and the contaminant specificities. The disposal of massive quantities of plant biomass materials polluted with heavy metals accumulated during the remediation process, notably phytoextraction [10, 11], is a significant impediment to the large-scale adoption of the technique. When the pollutant concentration in plant biomass exceeds a specific level, it is classified as a potentially hazardous object that must be stored or disposed of carefully [12]. An integrated solution was presented to solve this disposal challenge that included post-processing of biomass from phytoremediation plants to recover energy and high-value components [1]. It was concluded that improving post-process energy and elements from biomass will considerably enhance the financial sustainability of phytoremediation projects while also reducing the environmental implications of contaminated biomass disposal [1]. This highlights the value of integrating combustion with other thermochemical conversion processes, including gasification, pyrolysis, and biomass hydrothermal processing. This chapter will discuss thermochemical conversion technologies' fundamental principles and terminology, focusing on combustion and combustion technologies for a long-term energy and element recovery plan.

9.2 Biomass Fuels Contaminated with Heavy Metals

Apart from biomass material harvested from plants contaminated with heavy metals, there are various types of biomass materials (such as demolition wood, de-inking sludge from paper recycling, and sewage sludge) with high heavy metal content that can be used as fuels for bioenergy applications

Table 9.1 Some contents of heavy metal in various types of biomass fuels. Adapted from [13]

Biomass	Wheat Straw	Beech Wood	Demolish Timber	Phyto Remediation	Sewage Sludge	Chicken Litter	Paper Sludge
Metal content (mg kg^{-1} dry basis)							
As	0.18	3.5	550	22	(10)	–	8
Cd	0.2	1.0	8	–	38	–	<0.4
Co	3.0	–	–	–	–	–	9–12
Cr	25	2.5	1060	107	91	112	110
Cu	0.06	43	1080	70	330	71	310
Hg	6	0.12	10	8	2.7	–	1000
Mn	–	–	–	–	950	596	55
Ni		–	–	27	39	<10	–
Pb		33	6300	55	159	–	160
Zn		–	–	–	1318	209	470

and element recovery via combustion and other thermochemical conversion technologies due to seasonal variations in biomass fuels. Table 9.1 shows some heavy metal content levels for various biomass materials, adapted from a review on the fate of heavy metal during combustion and gasification of contaminated biomass [13], such as wood waste, demolition wood, phytoremediation biomass, sludges, de-inking sludge, and sewage sludge, and chicken litter. In addition, comparable statistics for two types of virgin biomass are supplied for comparison: wheat straw and beech wood. According to the literature, the numbers in Table 9.1 for demolition timber are mean values from a survey in the United States [1]. Furthermore, the phytoremediation biomass data is a composite of results from a study that looked at the performance of Amaranthus dubius at three extensively contaminated locations [2]. The information for sewage sludge is based on the average of samples collected across Europe [3]. All biomass fuels, except wheat straw and beechwood, have high levels of heavy metals, with demolition wood fuel being the worst, followed by sludges, chicken litter, and phytoremediation biomass fuels.

9.3 Combustion

9.3.1 Fundamentals of Solid Biomass Combustion

In general, the combustion of solid biomass fuels for heat and power generation may be defined as the rapid oxidation of the fuels, generating light and heat at considerable high temperatures. A typical combustion process is presented in Figure 9.1

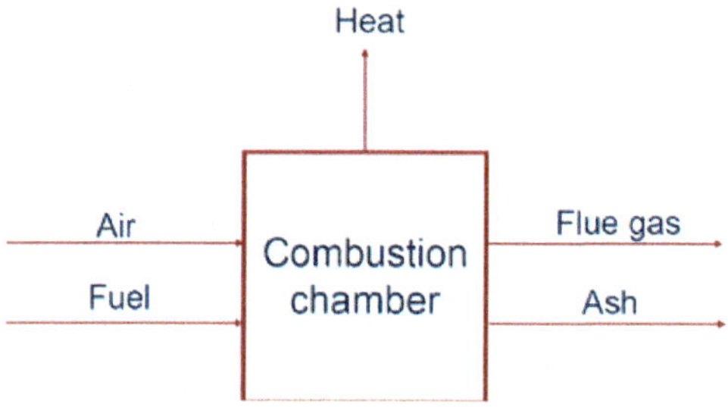

Figure 9.1 Typical combustion process

During combustion, chemical reactions between the major fuel constituents (mainly C and H_2) and oxygen (O_2) occur in the combustion chamber. Products resulting from the combustion process include flue gas, ash (incombustible mineral part of the fuel), and heat. Because biomass fuels are primarily composed of carbon, hydrogen, and oxygen, the main compositions of the flue gas are carbon dioxide and water, assuming complete combustion according to the two different mechanisms *combustion via hydroxylation* and *combustion via cracking*, presented in Figures 9.2 and 9.3, respectively.

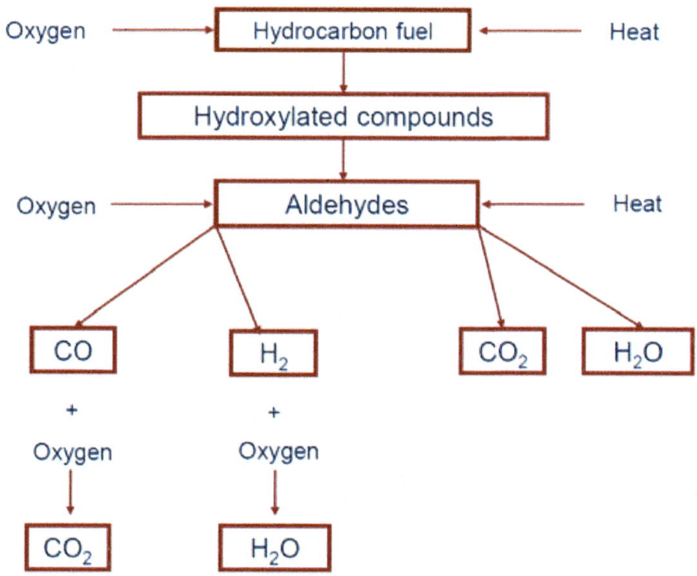

Figure 9.2 Combustion via hydroxylation.

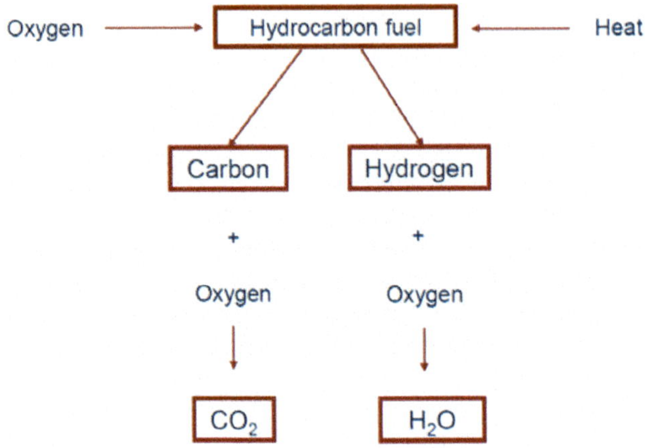

Figure 9.3 Combustion via cracking.

In virtually all practical combustion plants, both combustion mechanisms proceed simultaneously. Combustion via hydroxylation will predominate if the fuel-oxygen mixing and heating occur early and if the mixing time is extended. Explosion via cracking will prevail if the heating takes fast and the mixing time is short.

Usually, air with its oxygen content of 23% and 77% for N_2 by weight is employed as an oxygen carrier to provide the oxygen needed in combustion processes for heat and power generation. Therefore, the flue gas from air combustion typically contains a lot of nitrogen. In addition to these, fly ash, char, tar, soot particles, and other gaseous impurities including unburned hydrocarbons, dioxin, furans, VOCs (volatile organic compounds), NO_x, SO_x, and metals vapors (such as Hg, Se, As, and Cd) may also present in flue gas from combustion of biomass fuels.

The combustion of solid biomass involves a sequential series of stepwise processes, illustrated in Figure 9.4 [16]: *drying, devolatilization* or *pyrolysis, flame combustion, and char combustion*. Once introduced into the combustion chamber, solid biomass fuel is heated up, and the *drying* step takes place first (at around 100–105 °C). Then the *devolatilization* or *pyrolysis* step (at around 300–400 °C) proceeds, following the thermal front through the fuel particles and releasing volatile compounds. The volatile matters released from the devolatilization step include permanent gases (CO, CO_2, and CH_4) and condensable compounds, including water, methanol, acetic acid, acetone, and heavy hydrocarbons. The condensable organic compounds may condense to form a tarry liquid if cooled before burning. When moisture and volatile

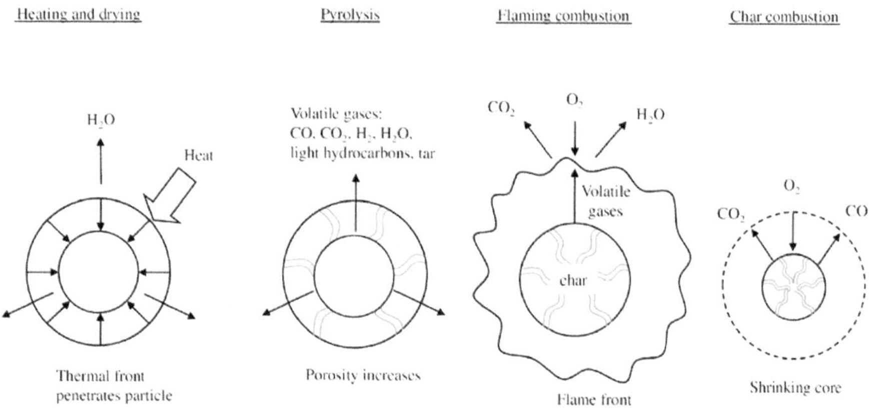

Figure 9.4 Processes for combustion of solid biomass fuels. Adapted from [16].

matters have gone, the combustible residues, i.e., porous char particles mainly consisting of fixed carbon and non-volatile ash constituents, are burning partly or entirely according to reactions (9.1) and (9.2), respectively.

- Combustion

$$C + 0.5\, O_2 = CO_2 \qquad -394\text{MJ/kmol} \tag{9.1}$$

- Partial combustion (Gasification with O_2)

$$C + 0.5\, O_2 = CO \qquad -111\text{MJ/kmol} \tag{9.2}$$

Among the four stages of burning solid biomass fuels, char combustion is the slowest and most restrictive; it is the quietest and is thus the most limiting. *Char combustion* is a heterogeneous gas-solid reaction that is governed by mass transfer of oxygen to the porous char particles' interior surfaces rather than chemical kinetics, which is typically extremely rapid at the increased temperature of burning. *Devolatilisation* is a highly quick process in comparison to the rest of the combustion process; it may take as little as a second for microscopic fuel particles but can take minutes for wood logs. It is a comparable endothermic process to pyrolysis (the anaerobic process by definition). This stage results in the release of a considerable quantity of volatile stuff into the gas phase, which prevents oxygen from the surrounding air from entering the thermal front where devolatilization occurs. Flame combustion is inextricably linked to devolatilization, which happens in the gas phase on the surface of the fuel particles when volatile substances are released into the gas phase and disperse into the surrounding air.

9.3.2 Fluidized Bed Combustion for Solid Biomass Fuels

Figure 9.5 shows different combustion technologies for burning solid biomass fuels, which are classified by the relationship between superficial gas velocity and pressure drop:

- Fixed bed combustion
- Fluidized bed combustion
- Pulverized fuel combustion

At low primary air velocity, the bed of particles is not mobilized and the gas penetrates upward through the bed. As the gas velocity increases, the pressure drops across the bed. When the pressure drop is equal to the weight of the bed, any increase in gas flow will cause expansion of the bed. When the gas velocity is more significant than *minimum fluidization velocity* or *fluidizing velocity* (W_L), gas bubbles rise through the bed and the bed is fluidized, called a *bubbling bed.* As the primary air velocity is much higher than the W_L, the bubbles pass the bed in a vigorous and undefined fashion. The turbulent fluidization occurs, a state of fluidization with very rapidly changing small voids in the bed. When the flow rate of the gas is higher than the *terminal velocity* of a single particle or *sinking velocity* (Ws), some of the particles

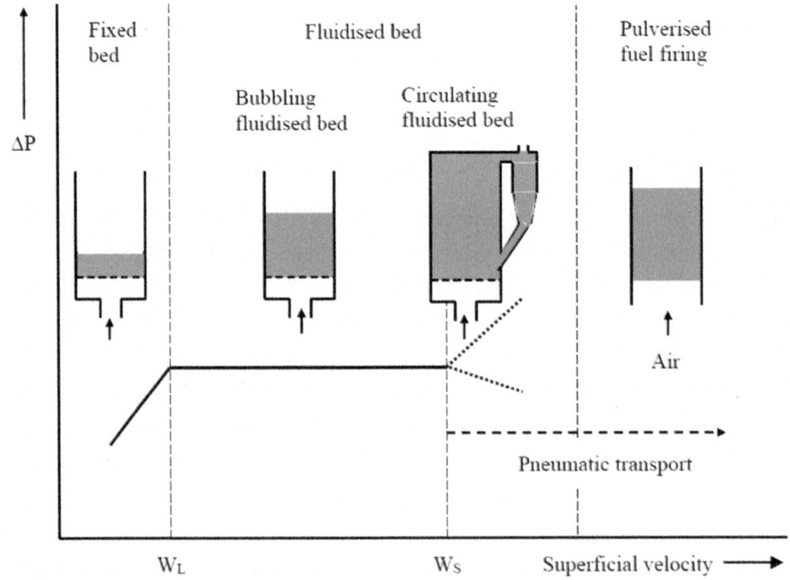

Figure 9.5 Combustion technologies for solid biomass fuels.

return to the bed while smaller particles will be carried over with the gas. It depends on, for instance, particle size, fluidization velocity, and the height of the combustion chamber. In the later fashion, the suspension of smaller particles extends itself over the entire space of the combustion chamber above the dense, bubbling bottom bed. Some of them may be elutriated out of the combustion chamber and thus must be re-circulated back to the fluidized bed to keep the solid mass in the bed constant.

The choice of combustion technology for burning biomass fuel is related to the fuel moisture content, density, heat value, physical form, ash content, the scale of operation, and the end-use of the energy. Among the combustion technologies available for firing solid biomass fuels, fluidized bed combustion (FBC) is suitable for large-scale applications due to its relatively high thermal efficiency and flexibility in fuel properties [3, 4].

The characteristic that distinguishes FBC is the suspension of a bed of hot particles in the main combustion air that is introduced from the bottom of the combustor. Because the bed acts similarly to a boiling fluid, it is said to be fluidized. According to the pressure at which combustion occurs, there are two types of FBC technology: atmospheric FBC (AFBC) and pressurized FBC (PFBC). Both AFBC and PFBC may be classified further into bubbling FBC (BFBC) and circulating FBC (CFBC), which are characterized by the surface gas velocity that defines the combustor's hydrodynamics.

Biomass may be burned efficiently using the FBC technology at temperatures as low as 800–900 °C, reducing slagging and fouling. The high concentration of alkali metals in biomass, on the other hand, might lead to particle aggregation in the bed material. Bed agglomeration during fluidized bed combustion of biomass is directly caused by the sticky coatings of Ca-K-silicates that cover the initial bed material. Combustion plant shutdowns and increasing operating expenses are a direct result of these issues. To absorb heavy and alkaline metals discharged into the gas phase during combustion, solid sorbents or getters may be added to the bed materials.

9.3.3 Ash Formation and Fate of Heavy Metals During Combustion of Solid Fuels

During the combustion of solid fuels, the organic matrix of the fuel particles is consumed, and metal species are subject to one of the transformation pathways described in Figure 9.6. The figure is valid for both coal and solid biomass fuels, except for the bimodal distribution being reasonable for coal fly ash only. In addition, it should be noted that coal fly ash is spherical particles but not biomass flash ash.

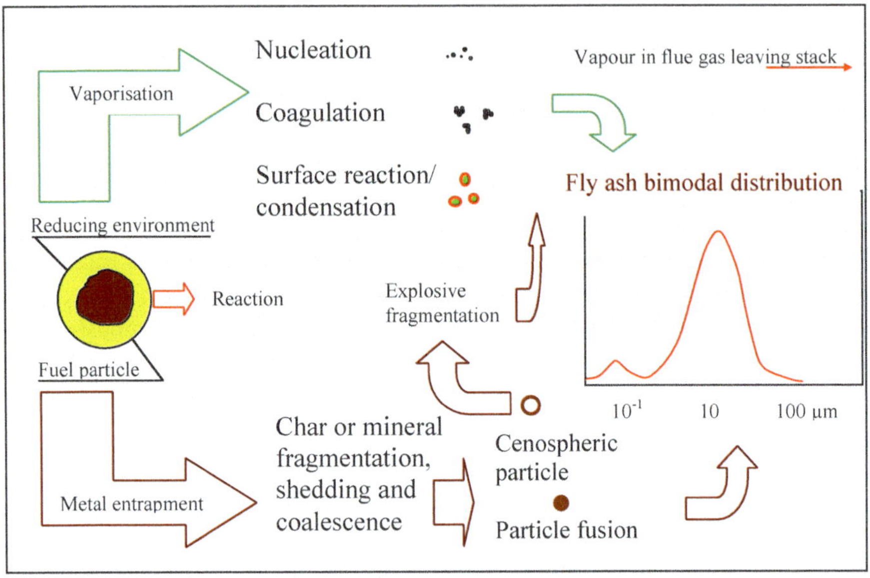

Figure 9.6 Metal transformations during combustion of solid fuels.

Once released, metal species may either react chemically on the surface of the ash-forming particles, with other reactive species in combustion gas or undergo physical transformation processes such as nucleation creating fume particles (0.2 μm). Nucleation may also be followed by coagulation or agglomeration processes producing coarser particles (10 μm). Tiny fume particles and part of volatile metals still in the gas phase are difficult to control and may leave the stack in the form of aerosol emission. Part of the inorganic metal species may be captured in char particles or react on the surface of bed material particles in the FBC boiler, forming silicate species. The species attached to the bed material will eventually be removed from the bottom ash from the bed as the material is periodically renewed. In contrast, the species attached to the fly ash will be collected in the particle control devices. Only a tiny part of the metals leaves the stack in gas form or/and small particles, together called aerosol emission.

The fate of heavy metal during the combustion of solid fuels have been studied based on theoretical calculation [19, 20] or experiments with coal [21–23], sewage sludge [24] or a mixture of these fuels [25], and woody or grassy plants [25–27]. The partitioning of heavy metals during the combustion of solid fuels, including biomass, can be classified into three broad

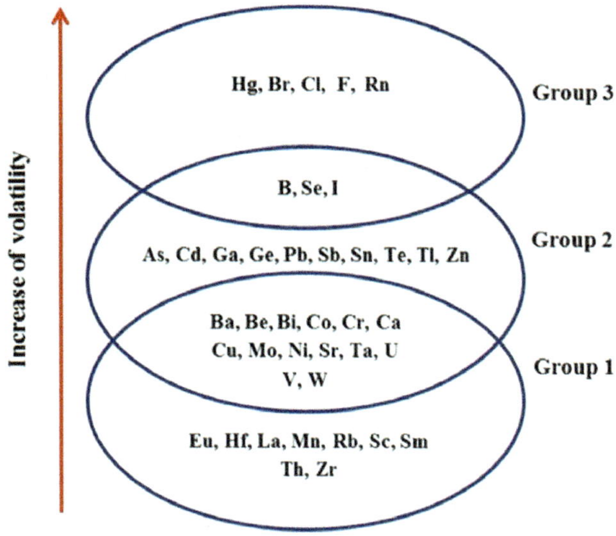

Figure 9.7 Metal distribution during combustion of solid fuels. Adapted from [28, 29].

groups according to their volatility, as presented in Figure 9.7 [28, 29]. Group 1 includes lithophile elements such as Ba, Mg, and Mn, concentrated in the bottom ash or partitioned equally between bottom ash and fly ash. Group 2 is represented by chalcophile elements such as Cd, As, Pb, Sb, and Zn, concentrated more in fly ash than bottom ash. They have even enriched the fraction of finer fly ash particles. The last group, Group 3 consists of elements mainly present in the gas phase, such as Hg, B, and Se. In addition to these, several elements may exhibit their partitioning behavior in between the groups. For example, Cr, Ni, U, and V may show partitioning behavior intermediate between Group 1 and Group 2, while B and Se are in between Group 2 and Group 3.

Due to the fact that metals in biomass are mostly found in aqueous solutions or are attached to organic structures, they are easily released after burning. Thus, rather than mineral transformations during coal combustion, the key mechanism involved in the primary generation of ash during biomass combustion is the vaporization of inorganic ash-forming species and the synthesis of ash minerals. Additionally, chlorine is a necessary component in ash production. It enhances the movement of a large number of inorganic chemicals during biomass burning. Chlorides are very stable at elevated temperatures, in the gas phase, and in the presence of metals. Numerous biomasses, such as straw and pulp sludge, contain close to 0.5 weight percent of chlorine. Chlorine concentration often has the same effect on the amount of metals evaporated during combustion as metal concentration does.

During combustion of biomass, alkali and heavy metals in the fuel may be released into the gas phase and leave the process or be concentrated in certain ash fractions by condensation, depending on the fuel characteristics, combustion temperature, and the gas atmosphere the combustion technology. Heavy metals such as Cd, Ni, Hg, Pb, and As are directly environmental and health concerns once released into the environment. Alkali metals and their compounds are responsible for material problems. In combustion plants, vaporized alkali species are likely to re-condense on the equipment surfaces in the forms of slagging and fouling, causing material damage and reduced ability to transfer heat.

9.3.4 Combustion Relevant for phytoremediation Plant Biomass Contaminated with Heavy Metals

When phytoremediation plant biomass is burned, heavy metals (HMs) that enter the combustion chamber with the fuels can either (1) remain as bottom ash in the combustion chamber, (2) results in the formation of solid particles in the flue gas (also called fly ash), or (3) both in the flue gas. After burning Cr-containing Leersia hexandra Swartz, the quantity of Cr in the bottom ash was observed to be $> 90\%$. According to Wen et al. [30], Cr levels in bottom fly ash, ash, and flue gas were equal to 11,2%, 77.6%, and 11.2% after 45 minutes of incineration at 800 °C. They also recommended keeping the combustion temperature below 800 °C to avoid Cr emissions into the atmosphere. After burning the Sedum plumbizincicola at 350 °C, Zhong et al. [31] discovered 70 per cent of the initial Cd in the fly ash. After the ryegrass was burned at 675 °C, Zhu et al. [32] exhibited a 64.4 per cent Cd rate of recovery in bottom ash. Pb and Zn from Sedum plumbizincicola were volatilized above 650 °C by Zhong et al. [31], followed by condensation below 650 °C. After burning Pteris vittata L. containing As, at 850 °C, Lei et al. [33] discovered As in the bottom ash. However, this amount only accounted for 9.0–23.5 per cent of the biomass feedstock's initial As content [33]. The greater the temperature, the faster the HM vaporizes, and more metals concentrate in fly ash and flue gas, whose concentrations are related to HM volatilization behavior [31]. Fly ash often contains more metal than is legally allowed, making it less suited for recycling [34]. The metals from the compounds that remained in the bottom ash were rapidly leached, limiting the bottom ash's use [35]. The Cd, Zn, and Pb that remained at the bottom were found to have low mobility [31]. According to Zhu et al. [32], higher incineration temperatures are advantageous in reducing Cd leaching capacity. The metal contents in the bottom ash were 1.5–3 times greater than in the Phragmites australis and Arundo donax tissues, according to Bonanno et al. [34]. On the other hand, metal-enriched ash had much lower metal amounts than the regulatory limitations for reusing ash in agriculture and forestry. As a result, biomass ash might be used as a fertilizer [34].

Some additives can prevent HMs from being transported into the gas phase after combustion in phytoremediation plant biomass. According to Wu et al. [36], kaolin removed 91.2 per cent of Cd and 88.1 per cent of Zn from the gas phase, while activated carbon removed 99.1 per cent Zn, 97.6 per cent Cd, and above 99 per cent of polycyclic aromatic hydrocarbons from the flue gas. The addition of kaolin in the flue gas can also lower NOx levels in the flue gas to values set by Chinese emission standards [36]. Wen et al. [30] burned the Leersia hexandra Swartz for 30 minutes at 800 °C after adding zeolite and CaO. Compared to the plant without additions, the Cr recovery rate in bottom ash increased by 12.2 per cent and 8.8 per cent, respectively. The addition of kaolin, montmorillonite, and -Al2O3 creates active Si- or Al-containing groups that bind Cd and prevent it from evaporating, lowering the Cd concentration in the leaching solution. Above 900 °C, montmorillonite was more effective than kaolin at retaining Cd [32]. The maintained As content in the bottom ash rose by 21.6–39.2 per cent in the combustion trials with CaO addition [33]. According to Jagodziska et al. [37], ammonium sulphate was successful in capturing Cr, Cu, and Hg, while halloysite connected with Cd, cobalt (Co), vanadium (V), and manganese (Mn) (Mn). The most effective Pb binding material was kaolinite.

The physical and chemical adsorption of the HMs is the principal mechanism of these additives in lowering fly ash quantities and limiting HM volatilization during incineration [30]. HMs react with these compounds at high temperatures, generating nonvolatile substances [38]. Zn and Cr, for example, were linked into nonvolatile silicates and aluminosilicates by zeolite [38]. The HMs can permeate into the structure of the additives due to eutectic melting on the additive surface [39]. Physical adsorption occurs when As interacts with CaO; however, this leads to an irreversible chemical reaction that produces Ca–As–O compounds such as $Ca_3(AsO_4)2$ and $Ca_2As_2O_7$ [40, 41]. Fly ash from incineration could readily clog and damage the flue gas collection system in real-world applications [38]. At the same time, the bottom ash that remains at the bottom of the furnace may cause corrosion. As a result, these disadvantages of incineration biowaste as a usage method must be addressed in the future.

9.4 Gasification

9.4.1 Gasification Fundamentals

Gasification may be defined as partial combustion or partial oxidation of (solid or even liquid) fuels, aiming at the production of combustible gases of carbon monoxide (CO), hydrogen (H_2), and methane (CH_4) as the main product. Apart from these combustible gases, non-combustible gases such as CO_2 and N_2 and other combustible gases such as H_2S, NH_3, and higher hydrocarbons may also present in the product gas mixtures, depending on the gasification agents and technologies employed. The mix of combustible and non-combustible gases produced from thermal gasification is called *producer gas, fuel gas, or product gas* (even city gas, town gas, or tail gas). After being conditioned and upgraded, the producer gas may be called *syngas*. Primary gasification agents are air, steam, and O_2, which would result in producer gases with heating values of approx. 4–7, 10–18, and 12–28 MJ/Nm^3, respectively.

Similar to combustion, the gasification of solid biomass involves a sequential series of stepwise processes, illustrated in Figure 9.8 [16]: *drying*, *devolatilisation* or *pyrolysis*, *char gasification*, and *gas-phase reactions*. The last step determines the final chemical composition of the producer gas. The *drying* and *devolatilization* or *pyrolysis* steps are similar to the direct

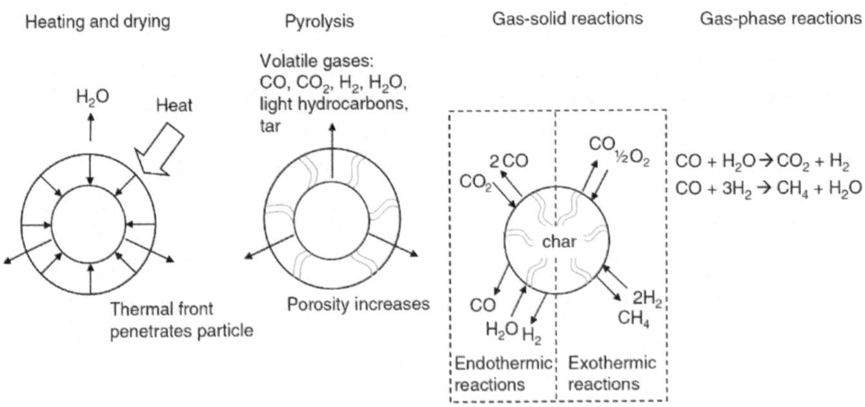

Figure 9.8 Processes for combustion of solid biomass fuels. Adapted from [16].

combustion presented in Figure 9.8 and the corresponding text. All kinds of combustion reactors can be employed for gasification. Note that for the fixed bed reactors, co-current contacting patterns between the fuel and gasification agent flows (downdraft entrained flow reactor, for example) are preferred to reduce tar formation.

The third step of biomass gasification is *char gasification*, which involves gas-solid reactions. Like indirect combustion, *char gasification* is the slowest and thus limiting step. The gas-solid reactions convert solid carbon in char into gaseous products, including CO, H_2, and CH_4 according to the reaction schemes below:

- Gasification with O_2

$$C + 0.5\,O_2 = CO \qquad -111\text{MJ/kmol} \tag{9.3}$$

- Gasification with CO_2 (Boudouard reaction)

$$C + CO_2 \leftrightarrow 2CO \qquad +172\text{MJ/kmol} \tag{9.4}$$

- Gasification with steam (Water-gas reaction)

$$C + H_2O \leftrightarrow CO + H_2 \qquad +131\text{MJ/kmol} \tag{9.5}$$

- Gasification with hydrogen (Hydrogasification)

$$C + 2H_2 \leftrightarrow CH_4 \qquad -75\text{MJ/kmol} \tag{9.6}$$

Among these reactions, Reaction (9.3) and Reaction (9.6) are exothermic, which together with the highly exothermic combustion reaction C + O_2 = CO_2 + (-394 MJ/kmol) contribute to supplying thermal energy required for drying, devolatilization, and the endothermic reactions (Reactions 9.4 and 9.5). In addition to this, the rate of Reaction (9.3) is the fastest among the four reactions of *char gasification*, which is followed by the rate of Reaction (9.5) being 3 to 5 orders of magnitude slower than that of Reaction (9.3). The char - CO_2 reaction (9.4) is approx. 6 to 7 orders of magnitude slower than that of the char–oxygen reaction (9.3). The char - H_2 reaction that forms methane (9.6) is the slowest.

The last step of biomass gasification involves gas-phase reactions (9.7 and 9.8), which determine the final chemical composition of the producer gas.

- Water-gas shift reaction

$$CO + H_2O \leftrightarrow H_2 + CO_2 \qquad -41\text{MJ/kmol} \tag{9.7}$$

- Methanation reaction

$$CO + 3H_2 \leftrightarrow CH_4 + H_2O - 206.1\text{MJ/kmol} \tag{9.8}$$

9.4.2 Gasification Relevant for Phytoremediation Plant Biomass Contaminated with Heavy Metals

Gasification's ability to safely transport and change heavy metals (HMs) throughout the remediation process, as well as provide clean producer gas with the desired composition and quality, needs more investigation [42]. Transformation of As in P. vittata increased as gasification temperature rose to 600 °C, with around 60% of As being liberated at this point in the process, according to Duan and colleagues [43]. Lin et al. [44] investigated the impact on HM volatilization and H2 production by adding silica sand, zeolite, calcium oxide, calcined coal, and activated carbon to the fluidized-bed gasification process. While H2 was detected in 53.1 percent of the gas created by activated carbon, H2 was the most common contaminant. Various cleanup facilities were gasified using the entrainment flow approach [45] by Jiang et al. Temperatures below 1000 °C made Cd, Pb, Zn, and As very volatile, although Mn and Cu were not [45]. Volatilization of HMs and their compounds is influenced by a variety of factors, including the temperature of the reaction, the velocity of the reaction gas, the bed material type, the particle size, and the reaction pressure [45–47]. It was recently reported that a microwave-assisted chemical looping gasification (MACLG) technique might produce syngas from hyperaccumulator biomass. Hematite, which has a high concentration of O, boosted H_2 and CO generation during MACLG of S. plumbizincicola and P. vittata [48]. The activated carbon was able to remove the HMs from the leftovers, although they were still present [48]. Calcium-based ($CaSO_4$) oxygen carriers with SiO_2 and Al2O3 as inert carriers may reduce the volatilization concentration of HMs in the gasification process, according to Chen et al. [49]. (Cd, Pb, Zn). The gasification process may waste energy, impair efficiency, corrode equipment, and clog pipelines if there is too much tar dust [50].

9.5 Pyrolysis

9.5.1 Pyrolysis Fundamentals

Biomass pyrolysis is a very complicated process, Figure 9.9, which may be defined as the thermal decomposition of biomass materials in the absence of oxygen or oxidizer, producing a wide range of products in liquid (tars, heavier hydrocarbons, and water), solid (charcoal or biochar) and gas (e.g., CO_2, H_2O, CO, C_2H_2, C_2H_4, C_2H_6, C_6H_6) phases. In practice, due to the economy of pyrolysis, the process may be carried out in the limited presence

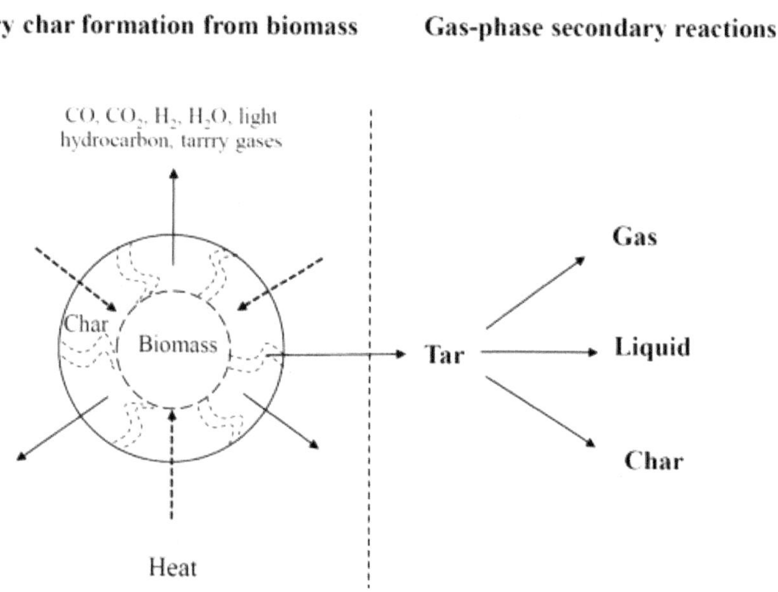

Figure 9.9 Process of biomass pyrolysis. Adapted from [51].

of oxygen or oxidizer such as flue gas. The distribution of pyrolysis products among the three physical forms depends on various factors, of which the most important are pyrolysis temperature and heating rate. If the heating rate is low (as low as <10K/min), the pyrolysis is called *slow pyrolysis*, resulting in a solid product (*charcoal* or *biochar*) as the main product. Slow pyrolysis usually is carried out at relatively low temperatures within 200–650 °C. If the heating rate is high (as high as 1000–10,000K/s), as the pyrolysis is called *fast pyrolysis*, resulting in a liquid product (*pyrolysis oil* or *bio-crude*) the main product. In addition to the solid and liquid products, some products in the gas form may also be formed via secondary transformation of tar as shown in Figure 9.9 [51].

Pyrolysis oil is a dark tarry liquid that contains up to 20% water [51]. Small molecules from lignin degradation are found in the continuous phase, which is an aqueous solution containing the products of decomposition. The discontinuous phase is mostly constituted of macromolecules of pyrolytic lignin [51]. Although there are several hundred chemicals in pyrolysis oil, the most fall into the following five general categories: Hydroxyaldehydes, Hydroxyketones, Sucrose, and dehydrosucrose, Carboxylic acids, and Phenolic compounds This makes pyrolysis a viable biorefinery process.

There are two process applications of slow pyrolysis, classified as *torrefaction* (200–300 °C) and carbonization (300–650 °C). The two processes are very similar, differentiated by the operating temperature ranges. In addition, the solid product resulting from torrefaction has lower carbon content than that from carbonization. More importantly, while the torrefied solid biomass can only be used as fuel for heat and power generation, biochar from carbonization can be used for various applications such as soil conditioners and fertilizer. It can also be activated to produce activated carbon for use as sorbent, catalyst carrier, and even for producing electrodes of supercapacitors.

9.5.2 Pyrolysis Relevant for Phytoremediation Plant Biomass Contaminated with Heavy Metals

When phytoremediation plant biomass materials are pyrolyzed, HMs enter biochar, bio-oil, and pyrolysis gas, and their distribution is remarkably similar to that of imprisoned biomass, as was the case with Cr-containing pyrolyzed plants that were also burned at 350 °C for comparison [30]. When the pyrolysis temperature was increased, the Cr concentration in the bottom ash increased in comparison to its burnt equivalent [30]. In spite of this, the Cr content of the gas generated during biomass pyrolysis was lower than that of the residue obtained by burning the same biomass at 350 °C [30]. Biochar, bio-oil, and gas formed from hyperaccumulators should be tested for their environmental safety before further use because of their high toxicity to humans and the environment. Numerous studies [52, 53] discovered that pyrolyzed biomass residues offered a decreased risk of HM leaching.

Additionally, the hyperaccumulator-derived biochar often contains fewer HMs, which is favorable for its safe disposal [52, 53]. Temperature, duration, additive, and entrained flow during pyrolysis must be optimized to further control these biochar risk variables [52, 53]. Increasing the pyrolysis temperature alone would not be sufficient to reduce the leachable content of several HMs in biochar [54, 55]. On the other hand, the optimal temperature is one of the most critical aspects in reducing the biochar's possible dangers. This optimal temperature may differ for different plants and HMs and their contents [56–59]. During slow pyrolysis, Chami et al. [60] recovered practically all Ni and Zn in the char. However, during rapid pyrolysis, they were only able to recover a few metals from the char. HM leaching from biochar can be reduced by adding additions (such as calcium dihydrogen phosphate, NaOH, $CaCO_3$, Al_2O_3, $FeCl_3$) [56, 61–63]. Some additives can react with HMs in the biomass and generate stable compounds during pyrolysis, as previously reported [54, 55]. Biochar has recently been shown to be helpful as an additive in the Phytoremediation of HM-contaminated soils, as it stimulates plant growth and improves the biochemical characteristics of these soils. However, more research is needed to determine the applicability and feasibility of this method (especially in the long run) [64–66]. HMs has accumulated in the bio-oil can be removed via cation exchange, solvent extraction, sorption, and other methods [67, 68]. The amount of moisture in the biomass impacts the yield and calorific value of the pyrolysis products, and it should be kept below 10% [35]. Pyrolysis-produced tar has the potential to clog pipelines, which can be costly to repair [35]. These disadvantages preclude pyrolysis from becoming widely used on an extensive basis. As a result, Phytoremediation based on pyrolysis is still in its early stages [35].

9.6 Hydrothermal Processing

9.6.1 Fundamentals of Hydrothermal Treatments of Biomass

Hydrothermal processing (HTP) utilizes hot compressed water (up to 700 °C and 40MPa) as the reaction medium, eliminating the need to dry the feedstock. HTL may be used to manufacture chemicals and biofuels in a variety of ways (Figure 6.1). The primary uses are as follows: 1) hydrothermal carbonization (HTC, 150–300 °C) for the production of solid fuels known as hydrochar; 2) hydrothermal liquefaction (HTL, 300–374 °C) for the production of a liquid fuel known as biocrude or bio-oil; and 3) supercritical water gasification (SWG, more than 374 °C) for the production of fuel gases as the primary product. Hydrothermal hydrolysis (below 180 °C) and wet torrefaction (180–250 °C) are two further uses of HTP, which are sometimes misunderstood since there is no clear-cut distinction in definition between the two ideas concerning operating temperature. However, whereas wet torrefaction creates advanced solid fuels only for energy consumption, HTC produces hydrochar that is suitable for use as a fuel, activated carbon, fertilizer, and soil conditioner.

The dielectric constant of water at 25 °C and 0.1MPa is 78, making it a polar solvent that is not suited for organic molecules that are predominantly nonpolar. Nonetheless, water works as a nonpolar solvent around the critical point of 374 °C and 22.1MPa, a very efficient reaction medium for organics. Due to the fact that water has a lower density in the liquid phase as it becomes hotter. As water's solvation power, hydrogen bond strength, polarity, dielectric strength, molecular diffusivity, and viscosity fluctuate, so does its density. The dielectric constant of water, for example, decreases considerably

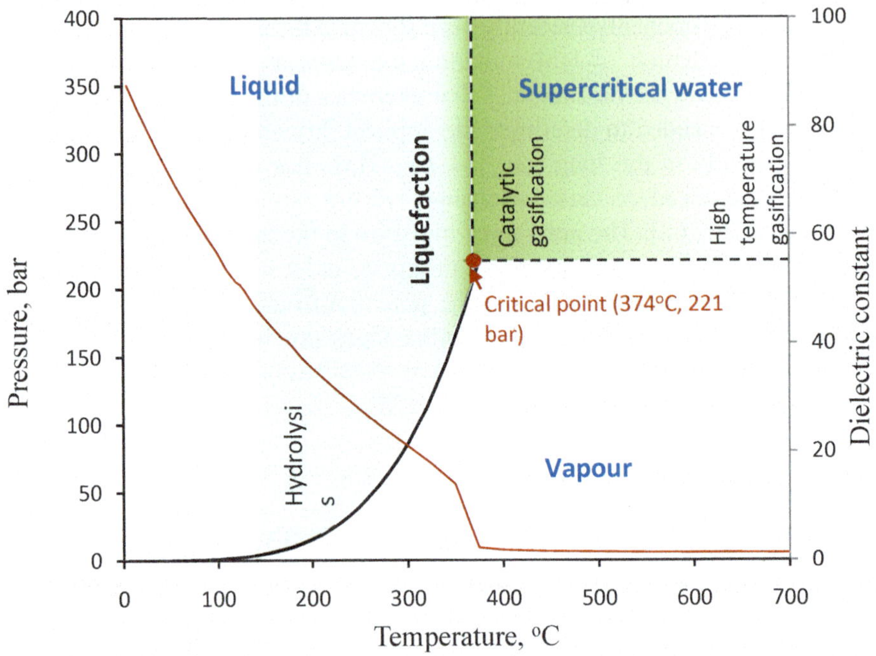

Figure 9.10 Pressure with temperature phase diagram of water and static dielectric constant of water at 200 bars as a function of temperature.

with temperature. Water at 200 °C has dielectric properties comparable to ambient methanol, water at 300 °C has dielectric properties similar to ambient acetone, water at 370 °C has dielectric properties equivalent to methylene chloride, and water at 50 °C has dielectric properties same to ambient hexane The ion product of hot compressed water is another intriguing characteristic. Under saturation vapor pressure, the ion product may reach a maximum temperature of roughly 250 °C. It begins to fall rapidly as soon as it passes the crucial point. Depending on the temperature and pressure, the hydrothermal medium may support ionic or free-radical processes. This occurs when densities rise beyond critical temperature and pressures are high enough to produce supercritical water (SCW). Free-radical reactions are more efficient at high temperatures and low densities. The production of char is inhibited by higher temperatures, although the presence of free radicals encourages the creation of coke. Solubility in ambient and hot compressed water is highly different between organics and non-organic substances like inorganics. Because of this difference, ambient water has a much higher concentration of inorganic chemicals dissolved than supercritical water, as seen in Table 9.2.

Table 9.2 Some physical properties of ambient and supercritical water.

Properties	Ambient water	Supercritical water
Dielectric constant	78	< 5
Solubility of organic compounds	Very low	Fully miscible
Solubility of inorganic compounds	Very high	≈ 0
Density ($g.cm^{-3}$)	1	0.2-0.9

9.6.2 Hydrothermal Treatments Relevant for Phytoremediation Plant Biomass Contaminated with Heavy Metals

HTC can treat hyperaccumulator plants with high water content as part of a phytoremediation process. According To Lee And Park, after HTC treatment, Cd, Cu, Ni, Pb, and Zn concentrations in hydrochars were lower than in raw sunflower biomass [69]. Furthermore, because the HMs isolated from the sunflower biomass were not destroyed but instead transformed to irregular forms during HTC, they remained in the liquid phase [69]. However, much more research is needed to determine how HMs are transported and chemical conditions in the hydrochar and liquid phases. This information would be crucial in assessing if the hydrochar produced could be used. Zn reduced the degradation of the phenolic hydroxyl group during HTC of the matching biomass, according to Mai et al. [70], affecting the redox functionality of the resultant hydrochar. HTC has better economic possibilities because it uses less energy and does not require pre-drying. However, before it can be used, the raw hydrochar-slurry must be filtered and dried. Another disadvantage is that HTC uses batch and semi-continuous settings, both of which must be changed for large-scale and commercialization. The salt produced during the chemical process is easily precipitated and has low solubility in subcritical or supercritical water [35]. When these residues interact with coke, the reactor can get blocked, raising the equipment's operating costs [35]. HTL is carried out in subcritical water at temperatures of 190–380 °C and pressures of 5–20 MPa [42]. Hydrochar and an organic liquid, commonly known as "bio-oil," are the end products [42, 71]. HTC reduces phytoremediation biomass more than incineration, pyrolysis, or gasification [35, 72].

HTL has been employed to separate HM from hyperaccumulators in a recent study [71, 73]. These efforts were accompanied by the reclamation of HMs, bio-oil, and hydrochar. More than 97 percent of Mn and other harmful metals in stems and leaves of Phytolacca americana L. can be separated into an aqueous solution when the particle size, liquid-solid ratio, pressure, temperature, reaction time, and catalyst are 75 m, 13.3: 1, 23 Mpa, 373 °C, 30 min, and 0.1 mol/L K_2CO_3, and 86.24 percent of biomass can be converted into crude bio-oil [73]. According to Qian et al. [71], around 90% of Zn was released from the S. plumbizincicola biomass during HTL at an optimal temperature of 220 °C, and the risk of release was decreased by using the HTL reaction to produce hydrochar. However, further research is needed to determine how HMs are dispersed in the hydrochar and the chemical form in the liquid phase. Compared to similar fuels made from pyrolyzed biomass, HTL-derived liquid fuel has greater heating values, lower oxygen, and higher carbon content [74]. Bio-oils cannot be utilized as a transportation fuel directly. They can, however, be blended or enhanced before being utilized to make diesel-based fuels. Upgrading bio-oils to produce liquid fuel can be done catalytically, as discussed and detailed elsewhere [75, 76]. The presence of HM (e.g., Zn, Pb, and Ni) in the hyperaccumulators before these bio-oils has also been reported to catalyze the hydrogenation reaction and the generation of organic acids from these bio-oils as beginning materials [77]. The in-situ up-gradation of the bio-oil resulted from the HTL of Zn/Pb/Ni-hyperaccumulator, especially if the released HM compounds could be identified and/or recovered. HTL, like HTC, has an issue with obstructing the reactor and increasing the equipment's operating costs [35].

HTG decomposes biomass in the presence of supercritical water at pressures and temperatures typically exceeding 22.1 MPa and 374 °C, respectively, and is compatible with wet biomass [42]. Reaction time is less effective than reaction temperature, and pressure has the least effect on conversion efficiency and product yields [78]. H2 and CH_4 are the most common gas products [42]. After incineration, pyrolysis, and gasification, the decrease of phytoremediation biomass processed by HTC is reduced [35, 72].

During HTG, supercritical water serves as both a medium and a reactant. HTG can treat fresh hyperaccumulator biomass without drying it [42], and it is particularly effective at reducing the environmental dangers of HM seeping from the hyperaccumulator while also recovering energy [68, 78]. The solubility of HMs in supercritical water during HTG is lower than in subcritical water during HTC [79], which is advantageous for treating hyperaccumulator-derived liquid products. There are three sub-categories of HTG [80]. The aqueous phase reforms at 215–265 °C with H2 and CO2 as final products falling into the first group [81]. The second category includes near-critical catalyzed gasification events that produce CH4 at 350–400 °C [82]. The third group comprises processes in which supercritical water gasification occurs at temperatures exceeding 500 °C, with the principal products being H2 and CO2 [83]. Unlike other gasification methods, HTG produces more H_2 from biomass regardless of its moisture level.

Furthermore, the creation of high-pressure H2 lowers the cost of gas compression, which is generally incurred during the storage of released H2 [84]. Because of its moderate temperatures, HTG inhibits HM volatilization, unlike conventional gasification [85]. In the gas generated during the HTG of Pteris vittata L., for example, hardly any HMs were found [68, 78]. Although Pb and Zn tended to accumulate in solid residues with a maximum increase of about 50% in total content, they were mostly converted to more stable oxidizable and residual fractions, which significantly reduced ecotoxicity and bioavailability with no noticeable increase in direct toxicity fractions [78]. Nonetheless, HTG's extremely harsh reaction conditions necessitate complex setups, raising fabrication and operation costs. Salt deposition and reactor corrosion are two further issues hindering HTG commercialization.

9.7 Conclusion and Perspective

The concept of combining post-processing of biomass materials from phytoremediation plants for recovery of energy and high-value elements is a potential and promising pathway towards sustainable Phytoremediation via thermochemical conversions of HMs contaminated biomass into thermal energy, value-added chemicals, and HMs recovery.

Because of the high capacity of plant biomass reduction, combustion, gasification, and pyrolysis are ideal for the disposal and valorization of biomass plants contaminated with HMs. Because of the low degradation rate of plant biomass, hydrothermal treatments are suited for the disposal of lower grade and moist biomass plants. Because of the secondary contamination created by the transfer of HMs into the gaseous state, combustion and gasification are particularly suitable for the disposal of plants with high boiling point HMs (e.g. Ni, Cr, Cu). Plants with medium and high boiling point HMs are particularly well-suited for pyrolysis and hydrothermal modification (e.g. Pb, Cd, Zn, Ni, Cr, and Cu). By-products from all procedures can be recycled as resources. Each approach contains flaws that will be the subject of future research.

Although the movement and transformation of HMs in polluted phytoremediation plant biomass during heat treatment processes have been explored to a limited extent, further research into the evolutionary pathways of various types of volatile HMs (such as Hg) is needed. More investigation into HM movement and transformation mechanisms in contaminated phytoremediation plant biomass during heat treatment is required.

Currently, there is little study on the mechanism of extraction approach for phytoremediation wastes contaminated with HMs. As a result, more outstanding research into the inner workings of extraction agents and HMs is required. Furthermore, the extraction solution's treatment must be studied and investigated. Metals and some valuable chemicals may be removed and recycled, and this is an area where research is also worthwhile.

References

[1] Y. Jiang, M. Lei, L. Duan, P. Longhurst, Integrating phytoremediation with biomass valorisation and critical element recovery: A UK contaminated land perspective, Biomass and Bioenergy 83 (2015) 328-339.

[2] H.I. Gomes, Phytoremediation for bioenergy: challenges and opportunities, Environmental Technology Reviews 1(1) (2012) 59-66.

[3] D. Desjardins, N.J.B. Brereton, L. Marchand, J. Brisson, F.E. Pitre, M. Labrecque, Complementarity of three distinctive phytoremediation crops for multiple-trace element contaminated soil, Science of The Total Environment 610-611 (2018) 1428-1438.

[4] E.E. Agency, Progress in management of contaminated sites (CSI 015), Europe Environmental Assessment Agency, Kongan, 6DK-1050, Denmark [Internet] (2007).

[5] V.C. Pandey, O. Bajpai, N. Singh, Energy crops in sustainable phytoremediation, Renewable and Sustainable Energy Reviews 54 (2016) 58-73.

[6] X. Liu, Q. Song, Y. Tang, W. Li, J. Xu, J. Wu, F. Wang, P.C. Brookes, Human health risk assessment of heavy metals in soil-vegetable system: A multi-medium analysis, Science of The Total Environment 463-464 (2013) 530-540.

[7] M. Summersgill, Remediation technology costs in the UK & Europe; drivers and changes from 2001 to 2005, Remediation technology costs in the UK & Europe; drivers and changes from 2001 to 2005 (2005) 1-8.

[8] D.J. Glass, Phytoremediation's economic potential, Phytoremediation toxic met. (2000) 15-31.

[9] A. Mojiri, L. Ziyang, R.M. Tajuddin, H. Farraji, N. Alifar, Co-treatment of landfill leachate and municipal wastewater using the ZELIAC/zeolite constructed wetland system, Journal of Environmental Management 166 (2016) 124-130.

[10] B. Dhir, S. Srivastava, Disposal of metal treated salvinia biomass in soil and its effect on growth and photosynthetic efficiency of wheat, International Journal of Phytoremediation 14(1) (2012) 24-34.

[11] A. Sas-Nowosielska, R. Kucharski, E. Małkowski, M. Pogrzeba, J.M. Kuperberg, K. Kryński, Phytoextraction crop disposal - An unsolved problem, Environmental Pollution 128(3) (2004) 373-379.

[12] M. Ghosh, S.P. Singh, A review on phytoremediation of heavy metals and utilization of its byproducts, Applied Ecology and Environmental Research 3(1) (2005) 1-18.

[13] A. Nzihou, B. Stanmore, The fate of heavy metals during combustion and gasification of contaminated biomass-A brief review, Journal of Hazardous Materials 256-257 (2013) 56-66.

[14] J.J. Mellem, H. Baijnath, B. Odhav, Translocation and accumulation of Cr, Hg, As, Pb, Cu and Ni by Amaranthus dubius (Amaranthaceae) from contaminated sites, Journal of Environmental Science and Health - Part A Toxic/Hazardous Substances and Environmental Engineering 44(6) (2009) 568-575.

[15] J. Werther, T. Ogada, Sewage sludge combustion, Progress in Energy and Combustion Science 25(1) (1999) 55-116.

[16] R.C. Brown, T.R. Brown, Biorenewable Resources: Engineering New Products from Agriculture: Second Edition, 2014.

[17] R. Van Den Broek, A. Faaij, A. Van Wijk, Biomass combustion for power generation, Biomass and Bioenergy 11(4) (1996) 271-281.

[18] I. Obernberger, Decentralized biomass combustion: state of the art and future development, Biomass and Bioenergy 14(1) (1998) 33-56.

[19] M. Díaz-Somoano, M.R. Martínez-Tarazona, Trace element evaporation during coal gasification based on a thermodynamic equilibrium calculation approach, Fuel 82(2) (2003) 137-145.

[20] S.Q. Turn, Chemical equilibrium prediction of potassium, sodium, and chlorine concentrations in the product gas from biomass gasification, Industrial and Engineering Chemistry Research 46(26) (2007) 8928-8937.

[21] R. Meij, Trace element behavior in coal-fired power plants, Fuel Processing Technology 39(1-3) (1994) 199-217.

[22] Y. Huang, B. Jin, Z. Zhong, R. Xiao, Z. Tang, H. Ren, Trace elements (Mn, Cr, Pb, Se, Zn, Cd 698 and Hg) in emissions from a pulverized coal boiler, Fuel Processing Technology 86(1) (2004) 23- 32.

[23] J.R. Bunt, F.B. Waanders, Trace element behaviour in the Sasol-Lurgi MK IV FBDB gasifier. Part 2 - The semi-volatile elements: Cu, Mo, Ni and Zn, Fuel 88(6) (2009) 961-969.

[24] G.P. Reed, N.P. Paterson, Y. Zhuo, D.R. Dugwell, R. Kandiyoti, Trace element distribution in sewage sludge gasification: Source and temperature effects, Energy and Fuels 19(1) (2005) 298- 304.

[25] E. Kurkela, P. Ståhlberg, J. Laatikainen-Luntama, Pressurised fluidised-bed gasification experiments with wood, peat and coal at VTT in 1991-1994: Part 2. Experiences from peat and coal gasification and hot gas filtration, VTT Publications (249) (1995) X-58.

[26] S.Q. Turn, C.M. Kinoshita, D.M. Ishimura, J. Zhou, The fate of inorganic constituents of biomass in fluidized bed gasification, Fuel 77(3) (1998) 135-146.

[27] D.C. Dayton, B.M. Jenkins, S.Q. Turn, R.R. Bakker, R.B. Williams, D. Belle-Oudry, L.M. Hill, Release of inorganic constituents from leached biomass during thermal conversion, Energy and Fuels 13(4) (1999) 860-870.

[28] B. Jagustyn, M. Kmieć, Ł. Smędowski, M. Sajdak, The content and emission factors of heavy metals in biomass used for energy purposes in the context of the requirements of international standards, Journal of the Energy Institute 90(5) (2017) 704-714.

[29] L.B. Clarke, The fate of trace elements during coal combustion and gasification: an overview, Fuel 72(6) (1993) 731-736.

[30] X.B. Wen, X.H. Zhang, J. Liu, A comparative study on the disposal of harvested products of Cr hyper - accumulator Leersia hexandra Swartz by incineration and pyrolysis, Indust. Saf. Environ. 19 Protect. 44 (2018) 73-77.

[31] D.X. Zhong, Z.P. Zhong, L.H. Wu, H. Xue, Z.W. Song, Y.M. Luo, Thermal characteristics and fate of heavy metals during thermal treatment of Sedum plumbizincicola, a zinc and cadmium hyperaccumulator, Fuel Processing Technology 131 (2015) 125-132.

[32] Z. Zhu, Y. Huang, J. Zha, M. Yu, X. Liu, H. Li, X. Zhu, Emission and Retention of Cadmium during the Combustion of Contaminated Biomass with Mineral Additives, Energy and Fuels 33(12) (2019) 12508-12517.

[33] G. Fang, Q. Wang, J. Zhou, Y. Lei, Z. Chen, Z. Wang, A. Pan, S. Liang, Metal Organic Framework-Templated Synthesis of Bimetallic Selenides with Rich Phase Boundaries for Sodium-Ion Storage and Oxygen Evolution Reaction, ACS Nano 13(5) (2019) 5635-5645.

[34] G. Bonanno, G.L. Cirelli, A. Toscano, R.L. Giudice, P. Pavone, Heavy metal content in ash of energy crops growing in sewage-contaminated natural wetlands: Potential applications in agriculture and forestry?, Science of the Total Environment 452-453 (2013) 349-354.

[35] M.J. Wang, G.D. Sheng, R. Wang, Progress in disposal technologies for plants polluted with heavy metals after phytoextraction, Journal of Agricultural Resources and Environment 38(2) (2021) 151-159.

[36] L. Wu, D. Zhong, Y. Du, S. Lu, D. Fu, Z. Li, X. Li, Y. Chi, Y. Luo, J. Yan, Emission and Control Characteristics for Incineration of Sedum Plumbizincicola Biomass in a Laboratory-Scale Entrained Flow Tube Furnace, International Journal of Phytoremediation 15(3) (2013) 219-231.

[37] K. Jagodzińska, K. Mroczek, K. Nowińska, K. Gołombek, S. Kalisz, The impact of additives on the retention of heavy metals in the bottom ash during RDF incineration, Energy 183 (2019) 854-868.

[38] X. Wang, Y. Huang, Z. Zhong, M. Niu, Y. Sun, Q. Zhang, Experimental study on the capture of heavy metals by in-furnace additives during MSW incineration, Zhongguo Dianji Gongcheng Xuebao (Proceedings of the Chinese Society of Electrical Engineering), Chinese Society for Electrical Engineering, 2012, pp. 15-21.

[39] T.K. Gale, J.O.L. Wendt, In-furnace capture of cadmium and other semi-volatile metals by sorbents, Proceedings of the Combustion Institute 30(2) (2005) 2999-3007.

[40] R.O. Sterling, J.J. Helble, Reaction of arsenic vapor species with fly ash compounds: Kinetics and speciation of the reaction with calcium silicates, Chemosphere 51(10) (2003) 1111-1119.

[41] Y. Li, H. Tong, Y. Zhuo, Y. Li, X. Xu, Simultaneous removal of SO2 and trace as2O 3 from flue gas: Mechanism, kinetics study, and effect of main gases on arsenic capture, Environmental Science and Technology 41(8) (2007) 2894-2900.

[42] X. Cui, J. Zhang, X. Wang, M. Pan, Q. Lin, K.Y. Khan, B. Yan, T. Li, Z. He, X. Yang, G. Chen, A review on the thermal treatment of heavy metal hyperaccumulator: Fates of heavy metals and generation of products, Journal of Hazardous Materials 405 (2021).

[43] L. Duan, X. Li, Y. Jiang, M. Lei, Z. Dong, P. Longhurst, Arsenic transformation behaviour during thermal decomposition of P. vittata, an arsenic hyperaccumulator, Journal of Analytical and Applied Pyrolysis 124 (2017) 584-591.

[44] C.L. Lin, M.H. Wu, W.C. Weng, Effect of the type of bed material in two-stage fluidized bed gasification reactors on hydrogen gas synthesis and heavy metal distribution, International Journal of Hydrogen Energy (2019) 5633-5639.
[45] Y. Jiang, A. Ameh, M. Lei, L. Duan, P. Longhurst, Solid-gaseous phase transformation of elemental contaminants during the gasification of biomass, Science of the Total Environment 563-564 (2016) 724-730.
[46] K. Froment, F. Defoort, C. Bertrand, J.M. Seiler, J. Berjonneau, J. Poirier, Thermodynamic equilibrium calculations of the volatilization and condensation of inorganics during wood gasification, Fuel 107 (2013) 269-281.
[47] M.H. Wu, C.L. Lin, W.Y. Zeng, Effect of waste incineration and gasification processes on heavy metal distribution, Fuel Processing Technology 125 (2014) 67-72.
[48] B. Zhang, J. Zhang, Z. Zhong, W. Wang, M. Zhu, Syngas production and trace element emissions from microwave-assisted chemical looping gasification of heavy metal hyperaccumulators, Science of the Total Environment 659 (2019) 612-620.
[49] T. Chen, B.S. Jin, X.J. Wang, Thermodynamic equilibrium analysis of heavy metal migration in chemical looping gasification of the phytoremediating plant, Acta Pet. Sin. Pet. Process. Sect. 36 20 (2020) 1239-1252.
[50] Y. Xiao, S. Xu, Y. Song, Y. Shan, C. Wang, G. Wang, Biomass steam gasification for hydrogen rich gas production in a decoupled dual loop gasification system, Fuel Processing Technology 165 (2017) 54-61.
[51] P. Basu, Chapter 5 - Pyrolysis, in: P. Basu (Ed.), Biomass Gasification, Pyrolysis and Torrefaction (Second Edition), Academic Press, Boston, 2013, pp. 147-176.
[52] X. Cui, Y. Shen, Q. Yang, S. Kawi, Z. He, X. Yang, C.H. Wang, Simultaneous syngas and biochar production during heavy metal separation from Cd/Zn hyperaccumulator (Sedum alfredii) by gasification, Chemical Engineering Journal 347 (2018) 543-551.
[53] S. Wang, B. Gao, Y. Li, Y.S. Ok, C. Shen, S. Xue, Biochar provides a safe and value-added solution for hyperaccumulating plant disposal: A case study of Phytolacca acinosa Roxb. (Phytolaccaceae), Chemosphere 178 (2017) 59-64.
[54] Z. Liu, B. Chen, L.A. Wang, O. Urbanovich, L. Nagorskaya, X. Li, L. Tang, A review on phytoremediation of mercury contaminated soils, Journal of Hazardous Materials 400 (2020).

[55] Z. Liu, L.A. Wang, H. Xiao, X. Guo, O. Urbanovich, L. Nagorskaya, X. Li, A review on control factors of pyrolysis technology for plants containing heavy metals, Ecotoxicology and Environmental Safety 191 (2020).

[56] L. Shi, L. Wang, T. Zhang, J. Li, X. Huang, J. Cai, J. Lü, Y. Wang, Reducing the bioavailability and leaching potential of lead in contaminated water hyacinth biomass by phosphate-assisted pyrolysis, Bioresource Technology 241 (2017) 908-914.

[57] S. Li, T. Zhang, J. Li, L. Shi, X. Zhu, J. Lü, Y. Li, Stabilization of Pb(II) accumulated in biomass through phosphate-pretreated pyrolysis at low temperatures, Journal of Hazardous Materials 324 (2017) 464-471.

[58] X. Gong, D. Huang, Y. Liu, G. Zeng, R. Wang, J. Wei, C. Huang, P. Xu, J. Wan, C. Zhang, Pyrolysis and reutilization of plant residues after phytoremediation of heavy metals contaminated sediments: For heavy metals stabilization and dye adsorption, Bioresource Technology 253 (2018) 64-71.

[59] J. He, V. Strezov, T. Kan, H. Weldekidan, S. Asumadu-Sarkodie, R. Kumar, Effect of temperature on heavy metal(loid) deportment during pyrolysis of Avicennia marina biomass obtained from phytoremediation, Bioresource Technology 278 (2019) 214-222.

[60] Z. Al Chami, N. Amer, K. Smets, J. Yperman, R. Carleer, S. Dumontet, J. Vangronsveld, Evaluation of flash and slow pyrolysis applied on heavy metal contaminated Sorghum bicolor shoots resulting from phytoremediation, biomass and bioenergy 63 (2014) 268-279.

[61] W.-J. Liu, W.-W. Li, H. Jiang, H.-Q. Yu, Fates of chemical elements in biomass during its pyrolysis, Chemical Reviews 117(9) (2017) 6367-6398.

[62] Z. Liu, B. Lu, B. He, X. Li, L.A. Wang, Effect of the pyrolysis duration and the addition of zeolite powder on the leaching toxicity of copper and cadmium in biochar produced from four different aquatic plants, Ecotoxicology and Environmental Safety 183 (2019).

[63] Y.N. Liu, Z.H. Guo, Y. Sun, W. Shi, Z.Y. Han, X.Y. Xiao, P. Zeng, Stabilization of heavy metals in biochar pyrolyzed from phytoremediated giant reed (Arundo donax) biomass, Transactions of Nonferrous Metals Society of China (English Edition) 27(3) (2017) 656-665.

[64] J. Paz-Ferreiro, H. Lu, S. Fu, A. Mendez, G. Gasco, Use of phytoremediation and biochar to remediate heavy metal polluted soils: a review, Solid Earth 5(1) (2014) 65-75.

[65] H. Lu, Z. Li, S. Fu, A. Méndez, G. Gascó, J. Paz-Ferreiro, Combining phytoextraction and biochar addition improves soil biochemical properties in a soil contaminated with Cd, Chemosphere 119 (2015) 209-216.

[66] G. Gasco, M.L. Álvarez, J. Paz-Ferreiro, A. Mendez, Combining phytoextraction by Brassica napus and biochar amendment for the remediation of a mining soil in Riotinto (Spain), Chemosphere 231 (2019) 562-570.

[67] J.G. Yang, C.B. Tang, J. He, S.H. Yang, M.T. Tang, Heavy metal removal and crude bio-oil upgrade from Sedum alfredii Hance harvest using hydrothermal upgrading, Journal of Hazardous Materials 179(1-3) (2010) 1037-1041.

[68] M. Carrier, A. Loppinet-Serani, C. Absalon, F. Marias, C. Aymonier, M. Mench, Conversion of fern (Pteris vittata L.) biomass from a phytoremediation trial in sub- and supercritical water conditions, Biomass and Bioenergy 35(2) (2011) 872-883. 21

[69] J. Lee, K.Y. Park, Conversion of heavy metal-containing biowaste from phytoremediation site to value-added solid fuel through hydrothermal carbonization, Environmental Pollution 269 (2021).

[70] D. Mai, R. Wen, W. Cao, B. Yuan, Y. Liu, Q. Liu, G. Qian, Effect of Heavy Metal (Zn) on Redox Property of Hydrochar Produced from Lignin, Cellulose, and d-Xylose, ACS Sustainable 834 Chemistry and Engineering 5(4) (2017) 3499-3508.

[71] F. Qian, X. Zhu, Y. Liu, Q. Shi, L. Wu, S. Zhang, J. Chen, Z.J. Ren, Influences of Temperature 836 and Metal on Subcritical Hydrothermal Liquefaction of Hyperaccumulator: Implications for the Recycling of Hazardous Hyperaccumulators, Environmental Science and Technology 52(4) (2018) 2225-2234.

[72] F.Z. Li, Y.T. Teng, Y.P. Zhang, Y. Liu, Research progress of disposal technology for heavy metal hyperaccumulator plants, Environ. Sci. Technol. 41 (2018) 213-220.

[73] Z. Deng, J. Yang, J. Li, X. Zhang, Removal of heavy metals and upgrading crude bio-oil from Phytolacca Americana L. harvest using hydrothermal upgrading process, Chinese Journal of Environmental Engineering 8(9) (2014) 3919-3926.

[74] G.W. Huber, S. Iborra, A. Corma, Synthesis of transportation fuels from biomass: Chemistry, catalysts, and engineering, Chemical Reviews 106(9) (2006) 4044-4098.

[75] K. Tekin, S. Karagöz, S. Bektaş, A review of hydrothermal biomass processing, Renewable and Sustainable Energy Reviews 40 (2014) 673-687.

[76] M. Kumar, A. Olajire Oyedun, A. Kumar, A review on the current status of various hydrothermal technologies on biomass feedstock, Renewable and Sustainable Energy Reviews 81 (2018) 1742- 1770.

[77] G. Yu, Y. Zhang, B. Guo, T. Funk, L. Schideman, Nutrient Flows and Quality of Bio-crude Oil Produced via Catalytic Hydrothermal Liquefaction of Low-Lipid Microalgae, Bioenergy Research 7(4) (2014) 1317-1328.

[78] J. Li, J. Chen, S. Chen, Supercritical water treatment of heavy metal and arsenic metalloid bioaccumulating-biomass, Ecotoxicology and Environmental Safety 157 (2018) 102-110.

[79] K. Sue, Y. Hakuta, R.L. Smith Jr, T. Adschiri, K. Arai, Solubility of lead(II) oxide and copper(II) oxide in subcritical and supercritical water, Journal of Chemical and Engineering Data 44(6) (1999) 1422-1426.

[80] H. Correa, G. Blanco-Wells, J. Barrena, A. Tacón, Self-organizing processes in urban green commons. The case of the Angachilla wetland, Valdivia-Chile, International Journal of the Commons 12(1) (2018) 573-595.

[81] D.J.M. de Vlieger, A.G. Chakinala, L. Lefferts, S.R.A. Kersten, K. Seshan, D.W.F. Brilman, Hydrogen from ethylene glycol by supercritical water reforming using noble and base metal catalysts, Applied Catalysis B: Environmental 111-112 (2012) 536-544.

[82] Y. Matsumura, T. Minowa, B. Potic, S.R.A. Kersten, W. Prins, W.P.M. Van Swaaij, B. Van De Beld, D.C. Elliott, G.G. Neuenschwander, A. Kruse, M.J. Antal Jr, Biomass gasification in near and super-critical water: Status and prospects, Biomass and Bioenergy 29(4) (2005) 269-292.

[83] A. Kruse, T. Henningsen, A. Smag, J. Pfeiffer, Biomass gasification in supercritical water: Influence of the dry matter content and the formation of phenols, Industrial and Engineering Chemistry Research 42(16) (2003) 3711-3717.

[84] A. Demirbas, Biorefineries: Current activities and future developments, Energy Conversion and Management 50(11) (2009) 2782-2801.

[85] H.J. Huang, X.Z. Yuan, The migration and transformation behaviors of heavy metals during the hydrothermal treatment of sewage sludge, Bioresource Technology 200 (2016) 991-998.

Index

About the Editors

Yashvir Singh, *Associate Professor, Department of Mechanical Engineering, Graphic Era Deemed to be University, Dehradun, Uttarakhand, India, 7895797503,* yashvirsingh21@gmail.com.
Post Doctoral Fellow, Faculty of Mechanical and Manufacturi ng Engineering, Universiti Tun Hussein Onn Malaysia, Parit Raja, Batu Pahat, Johor, Malaysia.

Yashvir Singh is presently working as a post-doctoral fellow at Univeristi Tun Hussein Onn Malaysia, Parit Raja, Batu Pahat, Johor, Malaysia. He is also associate professor in the Department of Mechanical Engineering, Graphic Era Deemed to be University, Dehradun, Uttarakhand, India. He has more than 15 years of teaching experience. He has written more than 130 research articles and published them in various peer-reviewed journals. He has published more than 15 chapters in books and editor of 2 books. He is also a reviewer and editorial board member of various journals. His specialization includes areas like tribology, biofuels, lubrication, manufacturing, Additive Manufacturing, etc. He has taught various subjects at UG and PG levels, such as Non-conventional Machining processes, advanced manufacturing processes, Additive Manufacturing, etc. He published various research papers in reputed journals and also reviewed various articles in the journals.

Prateek Negi, *Assistant Professor, Department of Mechanical Engineering, Graphic Era Deemed to be University, Dehradun, Uttarakhand, India, 8449542059,* prateeknegi060@gmail.com

Prateek Negi is currently associated as an assistant professor in the Department of Mechanical Engineering, Graphic Era Deemed to be University, Dehradun, Uttarakhand, India. His area of specialization includes Thermal Energy Storage, Energy Technologies, and Alternate Fuels and currently working on Tribological analysis of bio-lubricants. Being a reviewer of some peer-reviewed journals and having more than 20 journal and conference papers to his credit. Has taught various subjects at the UG level which include heat and mass transfer, thermodynamics, non-conventional energy resources, computer-aided design, and manufacturing, and operations research.

Wei Hsin Chen, *Distinguished Professor, Department of Aeronautics and Astronautics, National Cheng Kung University, Tainan, Taiwan,* weihsinchen@gmail.com, chewh@mail.nckuedu.tw, https://sites.google.com/gs.ncku.edu.tw/genfuelhttp://myweb.ncku.edu.tw/~chenwh/

Wei-Hsin Chen received his B.S. degree from the Department of Chemical Engineering, Tunghai University, Taichung City, Taiwan, in 1988, and a Ph.D. degree from the Institute of Aeronautics and Astronautics, National Cheng Kung University, in 1993. After receiving his Ph.D. degree, Dr. Chen worked in an iron and steel corporation as a process engineer for one and a half years (1994-1995). He joined the Department of Environmental Engineering and Science, Fooyin University in 1995 and was promoted to a Full Professor in 2001. In 2005, he moved to the Department of Marine Engineering, National Taiwan Ocean University. In 2007, he moved to Tainan as a Faculty Member at the Department of Greenergy Technology, National University of Tainan (2007-2013). He is a professor now (2013) at the Department of Aeronautics and Astronautics, National Cheng Kung University. Professor Chen visited Princeton University, USA, from 2004 to 2005, the University of New South Wales, Australia, in 2007, the University of Edinburg, UK, in 2009, and the University of British Columbia, Canada, from 2012 to 2013, as a Visiting Professor. His research topics include hydrogen production and purification, bioenergy, clean energy, carbon capture, and atmospheric science. He has published over 380 papers in international and domestic journals and conferences. He is also the author of four books concerning energy science and air pollution.